Built on Water

Floating Architecture + Design

Lisa Baker

Built on Water

Floating Architecture + Design

BRAUN

CONTENTS

- 6 — **PREFACE**
- 8 — **STEPHEN TURNER'S EXBURY EGG**
  PAD studio
- 14 — **FREISCHWIMMER**
  tun-architektur
- 18 — **THESAYBOAT**
  Marek Řídký
- 22 — **WATERVILLA DE HOEF**
  Waterstudio.NL
- 26 — **ARK HOTEL**
  Remistudio
- 30 — **FENNELL RESIDENCE**
  Robert Harvey Oshatz, Architect
- 36 — **LIVING ON THE EILBEK CANAL**
  DFZ Architekten
- 42 — **REM EILAND**
  Concrete
- 48 — **WINTERBADESCHIFF**
  Wilk-Salinas Architekten, Thomas Freiwald
- 54 — **FLOATING HOTEL**
  Baumhauer Eichler Architekten
- 58 — **ONE OF ONE**
  MONO Architekten, baubüro eins
- 62 — **WATERKANTHUS 100**
  büro13 architekten
- 66 — **DE OMVAL VILLA**
  +31architects
- 70 — **AMPHIBIOUS HOUSE**
  Baca Architects
- 74 — **AQUADOMI HOTEL**
  C.F. Møller Architects
- 78 — **IBA DOCK**
  Han Slawik Architekt /architech
- 84 — **WATERVILLA WEESPERZIJDE**
  +31architects
- 88 — **FLOATING RACETRACK**
  Baca Architects
- 92 — **PRINSENSTEIGER**
  oth_architecten
- 96 — **THE FLOATING EXPERIENCE**
  AMF Architekten Martin Förster
- 100 — **DRIJF IN LELYSTAD**
  Attika Architekten
- 106 — **HET BOSCH**
  Dreissen architects, Jager Janssen architects
- 112 — **THE CROWN, FLOATING VILLAGE**
  Baca Architects, Waterstudio.NL
- 116 — **GUERTIN BOATPORT**
  5468796 Architecture
- 120 — **FLOATING ECO HOUSE**
  Monika Wierzba
- 124 — **VIEWPOINT**
  Erkko Aarti, Arto Ollila & Mikki Ristola (AOR)
- 130 — **RADIO YEREVAN HOUSEBOAT ON THE EILBEK CANAL**
  sprenger von der lippe Architekten
- 134 — **FLOATING HOUSES HUMBOLDTINSEL**
  Baumhauer Eichler Architekten
- 138 — **FLOATING OFFICE FOR WATERNET**
  Attika Architekten
- 144 — **BROOKLYN BRIDGE PARK PIER 5**
  James Carpenter Design Associates
- 148 — **CAPSULE HOTEL**
  Denis Oudendijk

152 **TAFONI FLOATING HOME** Joanna Borek-Clement	196 **POINT ZERO** Nio Architecten	236 **KRAANSPOOR** oth_architecten
156 **ECOLODGE** Marijn Beije	200 **BALTIC SEA ART PARK** WXCA	240 **FLOATING HOUSES, AMSTERDAM** Architectenbureau Marlies Rohmer
160 **PAMPUS HAVEN** MVRDV	204 **HOUSEBOAT ON THE EILBEK CANAL** Rost.Niderehe Architekten I Ingenieure	246 **CITADEL** Waterstudio.NL
164 **TEA HOUSE – BAMBOO COURTYARD** HWCD	208 **RE:VILLA** WHIM Architecture	250 **XINJIN WATER CITY** MVRDV
168 **DOCKS – CITE DE LA MODE ET DU DESIGN** Jakob+MacFarlane	212 **HARVEST DOME 2.0** SLO Architecture	254 **ARCHIPELAGO CINEMA** Büro Ole Scheeren
174 **FLOATING CINEMA** Duggan Morris Architects	216 **AMILLARAH – FLOATING PRIVATE ISLANDS** Dutch Docklands International, Waterstudio.NL	258 **THE OCEAN FLOWER** Dutch Docklands Maldives, Waterstudio.NL
178 **THEMATIC PAVILION** ONL [Oosterhuis_Lénárd]	220 **SEOUL FLOATING ISLANDS** Haeahn Architecture + H Architecture	262 **BLOOMING BAMBOO HOUSE** H&P Architects
182 **FLOATING HOUSE, SAN FRANCISCO** Robert Nebolon Architects	224 **SAN FRANCISCO BAY HOUSE** Robert Nebolon Architects	266 **CUBE ORANGE** Jakob+MacFarlane
188 **TWIN BLADE** Nio Architecten	228 **AALBORG HAVNEBAD** JWH arkitekter	270 **INDEX**
192 **HAUSBOOT ON THE EILBEK CANAL** martinoff architekten	232 **IBA WATERHOUSES** Schenk+Waiblinger Architekten	

PREFACE

Living on the water has always been one of humanity's brightest dreams and visions. The 1922 reconstruction of the lake dwellings in Unteruhldingen on the Lake of Constance (Germany) are proof that people lived on the water as far back as the Stone and Bronze Ages. This way of life was relatively well protected from attacks from animals or enemies. Furthermore, such houses also made it possible to fish from home. However, the oldest and most famous prehistoric stilt dwellings were just built on boggy ground, on the banks of lakes and not in deep flowing water. This way of living appeared in cultures all over the globe. Pictures of Amsterdam and Venice in particular, or individual constructions such as West Pier in Brighton by Eugenius Birch (1866) or Vito Acconicis' "Murinsel" in Graz (2002), demonstrate just how much architecture both on and in water spurred on architectural development.

However, fear of the water is just as deeply ingrained and intercultural as love of water: Genesis, the Gilgamesh Epic and the Atrahasis Epic all speak of a universal flood capable of wiping out the human race. An arc offered a safe place of retreat from life-threatening floods and this age-old fear is also played out in numerous Hollywood films. The construction of Noah's Arc is detailed in the bible and is therefore a much-use motif in Christian art. Usually, numerous craftsmen are depicted aiding Noah with the construction. The smallest documented measurements results in a box-shaped construction six stories high and with exterior measurements of 135 x 22.5 x 13.5 meters. This looks more like a raft with a barn perched on top of it – as seen on Michelangelo's ceiling in the Sistine Chapel – than a real boat. Noah wasn't required to steer, just to float, so the addition of a hull was unnecessary. This raft was the same length as 1.3 football pitches, or put another way, it was roughly half as long as the Titanic.

The chance that the entire planet will flood is of course extremely small, but catastrophic floods have always presented humanity and architecture with an immense challenge: floods in Asia are reported on the news almost every year, but catastrophic flooding in The Netherlands in 1953 – the water rose 5.25 meters above its normal level – really goes to show that rising water levels can also pose a threat to Europe.

The myth of Atlantis – the city that sank into the sea – is somewhere between the two poles of fascination and fear. According to Plato, the City of Atlantis was located on the other side of the Pillars of Hercules, or the Strait of Gibraltar, on an island in the Atlantic Ocean and sank in around 9600 BC after a natural catastrophe. However, other ancient authors doubt the hypothesis that this nation ever existed. The "Sixth Continent" is a reappearing motif in art and culture, for example in Pierre Benoit's novel L'Atlantide in 1919, in the works of J. R. R. Tolkien or Joseph Beuys, or in the Marvel comics. Life under water is also an attractive alternative for some architects, inspired by, for example, the Nautilus by Jules Vernes. The underwater city Futurama II, sponsored by General Motors, was on display at the World Fair in New York in 1964. Just like the company's first Futurama for the World Fair in 1939 by Norman Bel Geddes, this was also envisaged as an option for the near future – although even now it remains an utopia. On the other hand, living on the water is a reality in many places: The Chao-Le, "sea people" from Thailand live in stilt dwellings, houseboat owners live in houseboats on the Amsterdam canals. In Amsterdam living in a houseboat on the canals has been legal since 1652 – although it has always been a point of contention whether these can till be classed as boats or whether they should be considered immobile buildings. Houseboats are no longer required to be able to propel themselves forwards, but they must be able to be towed to other waters at any time (at least in theory). This is a limiting factor in the construction of many bridges. The idea of living on water is becoming increasingly popular in many other locations. The International Architecture Exhibition in Hamburg also focused on this subject from 2007–2013, using the motto "Designs for Future Metropolises", even building one of the main buildings on water. Numerous cities enjoy swimming and bathing facilities that float on the water, an idea that has been around since the 19th and early 20th centuries, although only a few of these have been maintained (Vlatava-Bad in Prague by Josef Ondrej Kranner, 1840; Inselbad Stuttgart by Paul Bonatz & Scholer, 1927).

Floating constructions are beginning to take on a whole new dimension: hotels, cinemas, cultural institutions, offices, even a racetrack are all presented in this volume, next to individual water villas and entire residential estates. Such constructions always raise the questions of how they float: is it supported by pillars on the ground, is it raised on stilts or does it float on a pontoon. A range of innovative techniques is used – from recycled floats to concrete pontoons.

Created and designed by PAD studio and artist Stephen Turner, the Exbury Egg was inspired by the nesting seabirds on the shore. It was built locally, by boat builder Paul Baker, as a cold-molded cedar plywood sheathed structure and the artist will track the aging process. Local Douglas fir was used for the supporting ribs and internal framing; continuing the age-old tradition of timber marine construction, which can be traced back many centuries on the Beaulieu River. The intent of the project was to explore the creation of a minimal impact live/work structure, using materials with a low embodied energy sourced within a twenty-mile radius, and put together by a team of local craftsmen using centuries old techniques.

STEPHEN TURNER'S EXBURY EGG
HAMPSHIRE, UNITED KINGDOM

Architects: PAD studio
Location: Beaulieu River, Hampshire, United Kingdom
Completion: 2013
Client: Space Placemaking and Urban Design
Gross floor area: 22 sqm
Function: living
Built on water: floats on recycled plastic buoyancy tanks filled with water

FREISCHWIMMER
HAMBURG, GERMANY

This floating house is designed to meet the needs of city residents; the layout accommodates flexible rooms of a sensible size to suit their purpose. Essentially, the house comprises two cubes joined together and made of different materials; the timber frame construction is anchored to the concreted base. Entrance to the house is via the roof terrace. Inside, the main material used is wood; wooden floorboards and walls create a bright and welcoming atmosphere. The exterior is clad with corten steel and untreated wood. The use of corten steel reflects Hamburg's industrial roots, while the integration of porthole windows reinforces the connection to traditional ship construction. A regenerative energy concept uses the canal water as a heat source.

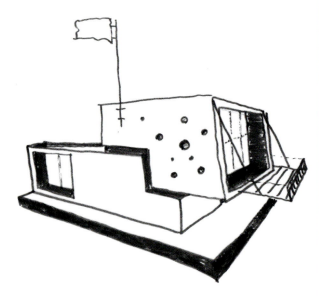

Architects: tun-architektur
Location: Uferstraße 8b, 22081 Hamburg, Germany
Completion: 2009
Client: confidential
Gross floor area: 120 sqm
Function: living
Built on water: floats on concrete pontoon

Ice has a lower density than water; therefore, it floats.

Architects: Marek Řídký
Structural engineers: Arrbo
Completion: 2012
Client: Marek Řídký
Gross floor area: 63 sqm
Function: living

THESAYBOAT
NELAHOZEVES, CZECH REPUBLIC

The construction of a minimalist house on the water was intended to provide comfortable year-round housing for two people or to be a family residence at the weekends. The houseboat is built with a clear vision that can be summarized in Corbusier's assertion that "human habitation should be a cell with a view of the stars". The house is designed as a highly functional space and borrows elements from both shipping and housing design. The use of wood cladding both inside and out reinforces the relationship between the house and surrounding nature.

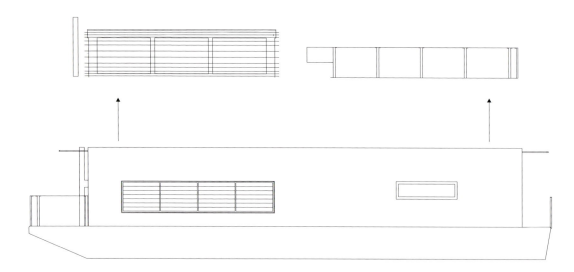

"Water is the mother of the vine, the nurse and fountain of fecundity, the adorner and refresher of the world." — Charles Mackay, Scottish writer

"Water is the mother of the vine, the nurse and fountain of fecundity, the adorner and refresher of the world." — Charles Mackay, Scottish writer

This spacious dwelling demonstrates that floating architecture has become a valid equivalent to land-based dwellings. Modern, light and transparent, this residence is an amphibious structure, floating on the water yet surrounded by land on three sides. The design places great emphasis on transparency and aims to emphasize the experience of living on the border between the land and the open sea. The large windows allow views over the garden and the water. Inside, all the spaces are connected, separated only by floor-to-ceiling cupboards that form the only fixed interior elements.

WATERVILLA DE HOEF
DE HOEF, THE NETHERLANDS

Architects: Waterstudio.NL
Location: Vinkenslag 11, 1426 AM, De Hoef, The Netherlands
Completion: 2006
Client: confidential
Gross floor area: 147 sqm
Function: living

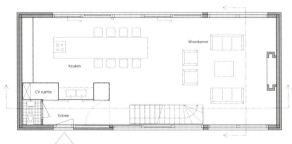

Atomic hydrogen has the lowest atomic mass of all chemical elements. It comprises one proton and one electron but almost never appears in this form in nature.

ARK HOTEL
SOCHI, RUSSIA

The Ark project was designed according to the concept of a bioclimatic house that can function independently. The building has its own energy system and can even produce energy for supplying nearby houses and powering green transport. The structure of the building allows it to float and exist autonomously on the surface of the water. All the plants are chosen according to principles of compatibility, illumination and efficient oxygen production. Light is drawn in through the transparent glass roof. The balconies serve as recreational space.

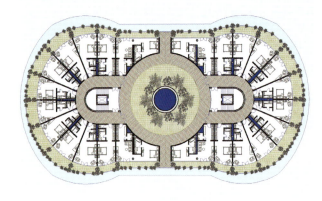

Architects: Remistudio
Location: Nizhneimeretinskaya Cove, Adler district, Sochi, Russia
Completion: future
Client: confidential
Gross floor area: 15,000 sqm
Function: hotel

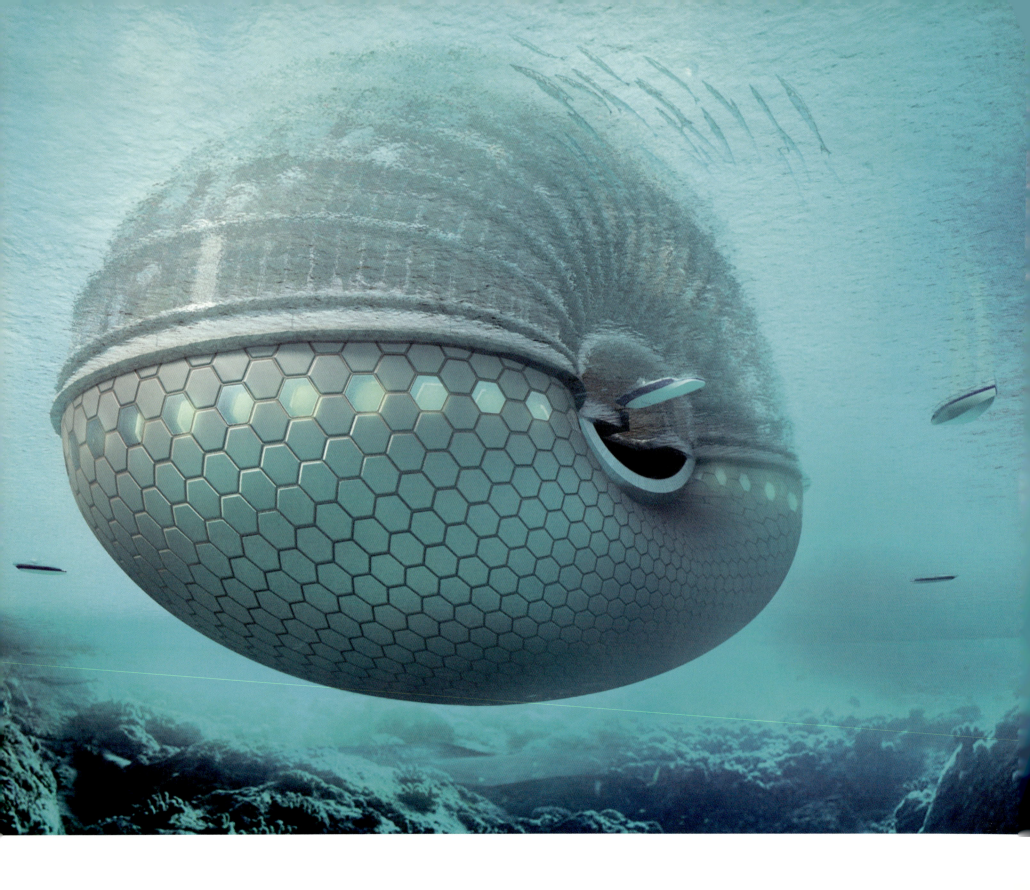

The Urus are a pre-Incan society that lives on a group of 42 floating island on Lake Titicaca, Peru.

Architect Robert Harvey Oshatz is known for creating architectural structures that are at peace with their environment. The unique qualities of the specific site and the client's "needs and wants" are the driving force behind each design idea. The Fennell house presented a unique challenge as the site was "on" the Willamette River as opposed to "by" the river. As the architect studied the project he focused on the poetry of the ripples and contours of the river, its never-ending flow, the view and the interrelationship concerning the play of the sun and moon as it courses through the days of the year. Curved glue lam beams were used to capture the timeless sense of flowing water and time passing and its relationship with the river, creating a spiritual and poetic sense of space.

FENNELL RESIDENCE
PORTLAND, OR, USA

Architects: Robert Harvey Oshatz, Architect
Structural engineers: Brett King
Location: 6901 SE Oak Park Way, Portland, OR 97202, USA
Completion: 2005
Client: Randy Fennell
Gross floor area: 220 sqm
Function: living
Built on water: supported by wooden logs, floats on styrofoam flotation cubes

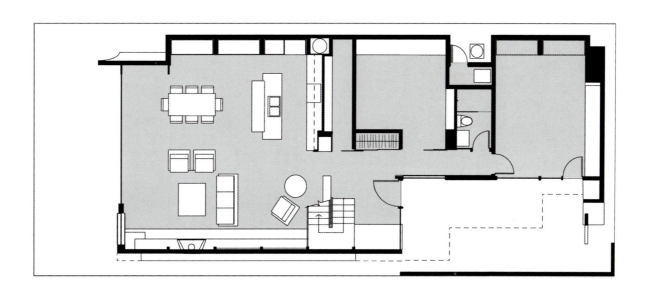

"Everything flows." –
Heraklit, Greek philosopher

LIVING ON THE EILBEK CANAL
HAMBURG, GERMANY

The design of this project is oriented around ship architecture and explores the subject of how to use space in the best possible way. All the rooms and surfaces in this house serve more than one function. The houseboat is structured into different volumes all connected to each other. The design as a whole incorporates a variety of different perspectives, both from land and from the water. The volumes vary in height and position, resulting in a wide variety of interior spaces. Facing the water, large windows determine the design of the house and frame the views across the water. The two volumes are separated by a narrow incision; this provides access and helps to draw light deep inside.

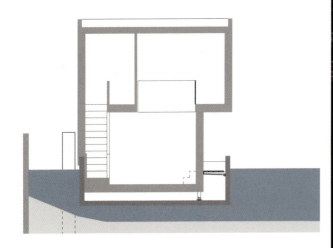

Architects: DFZ Architekten
Location: Uferstraße, 22081 Hamburg, Germany
Completion: 2010
Client: confidential
Gross floor area: 200 sqm
Function: living
Built on water: floats on reinforced concrete pontoon

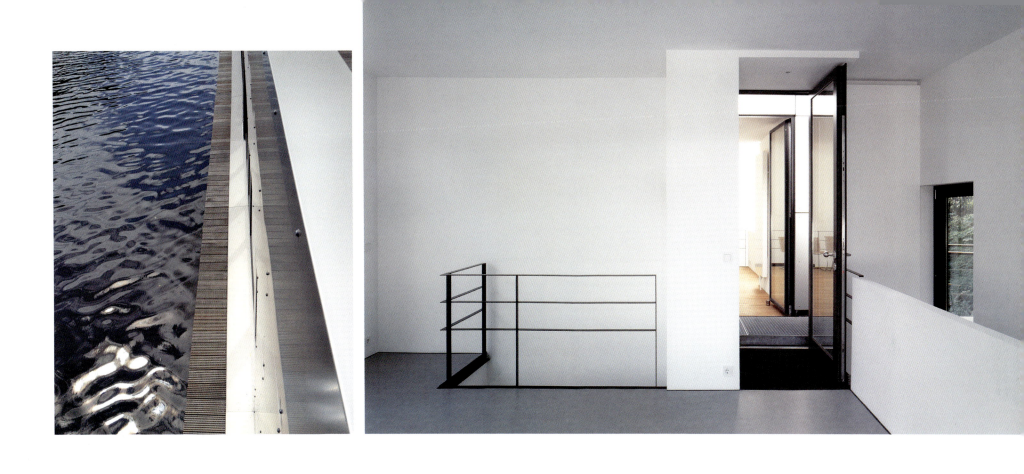
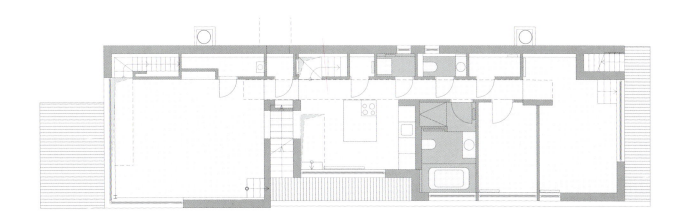

Beyond the tree line in the arctic, driftwood was the only source of wood available to the Inuit and other indigenous peoples.

Architects: Concrete
Structural engineers: ABT Delft
Location: Haparandadam 45, Amsterdam, The Netherlands
Completion: 2011
Client: De Principaal Amsterdam
Gross floor area: 661 sqm
Function: restaurant, office
Built on water: supported by steel columns

REM EILAND
AMSTERDAM, THE NETHERLANDS

In 2007 Concrete was asked by hospitality entrepreneur Nick van Loon to come up with an idea for REM-island located in the IJ in Amsterdam. The program consists of an office function on the first deck and a restaurant on decks two and three, an extra story has been created spreading the restaurant over two floors. The distinctive red and white-checkered building rests on 12-meter-high columns and is located 15 meters offshore. Walking around the island the viewer is treated to a great view from every terrace. The red and white-checkered pattern of the building is continued in the new extension. Long external footbridges, large signal lights and a lifeboat are also located on the island.

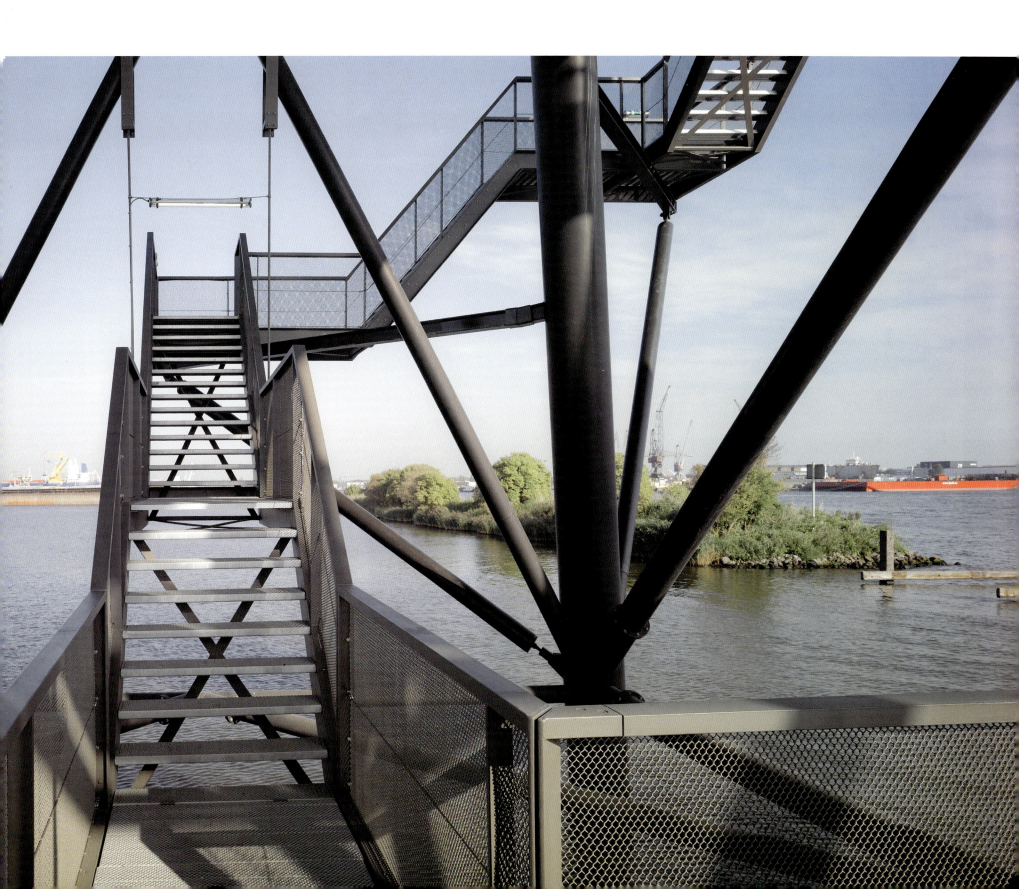

Fish that don't have a swim bladder but also don't count as ground-dwelling fish, the shark for example, have to propel themselves forward with continuous swimming.

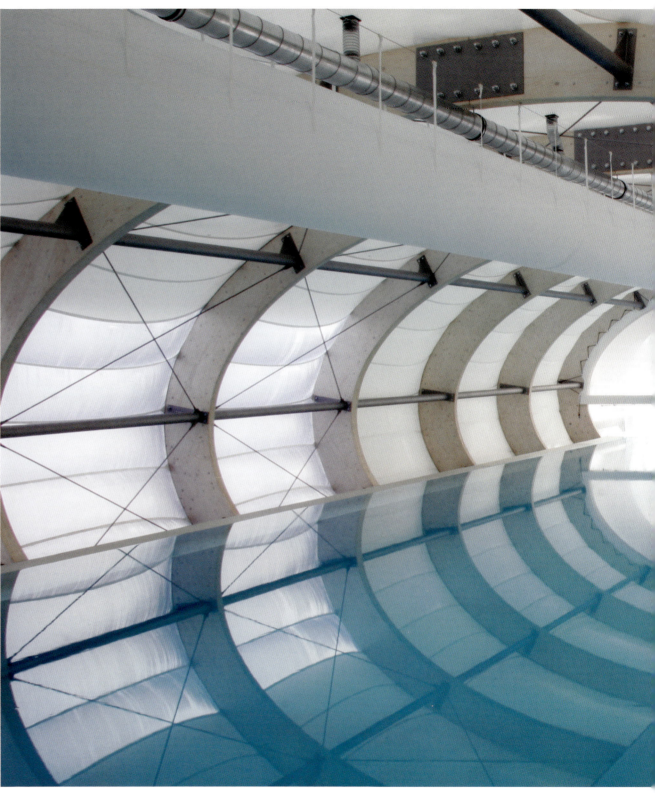

This location was once used by barges, before all use of the water surface was banned during the era of the Berlin wall. It is now home to a swimming pool ship that in winter also serves as a covered sauna in Berlin's east harbor. The sauna was built in addition to the swimming pool, designed by AMP arquitectos with Gil Wilk and artist Susanne Lorenz in 2004, which was originally conceived for summer use only. One year later, Wilk-Salinas Architekten created a covered pool area, sauna and café with a cozy room temperature of 25 degrees, the perfect location to get away from the cold winter temperatures. The membrane envelope allows ample daylight through to illuminate the interior and also allows light out at night, so that the construction shines out like a beacon in the harbor.

WINTERBADESCHIFF
BERLIN, GERMANY

Architects: Wilk-Salinas Architekten with Thomas Freiwald
Location: east harbor section of the River Spree, Berlin, Germany
Completion: 2005
Client: Kulturarena Veranstaltungs GmbH
Gross floor area: 1,000 sqm
Function: leisure
Built on water: pier construction resting on pillars, floating pool fixed by chains

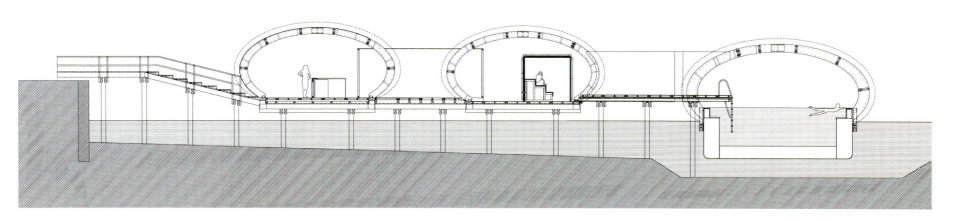

"The cure for anything is salt water – tears, sweat, or the sea." – Isak Dinesen (Karen Blixen), Danish writer

Architects: Baumhauer Eichler Architekten
Location: various
Completion: 2016
Client: confidential
Gross floor area: 4,423 sqm
Function: hotel
Built on water: floats on concrete pontoon

FLOATING HOTEL
VARIOUS

The concept for the floating hotel was based on the idea of being able to pull up directly at your hotel on your own landing pier after a sailing trip. The design is now being planned for a number of locations: as a floating spa hotel on a lake or as a resort for yacht sailors from the German North Sea to the Indian Ocean. For high sea locations, the plans involve a 60-by-60-meter floating platform of reinforced concrete and expanded polystyrene, always protected by a breakwater or located in a decommissioned harbor. The floating hotel will have 24 44-square-meter, two-story suites, as well as a number of restaurants and an outdoor area with pools and terraces.

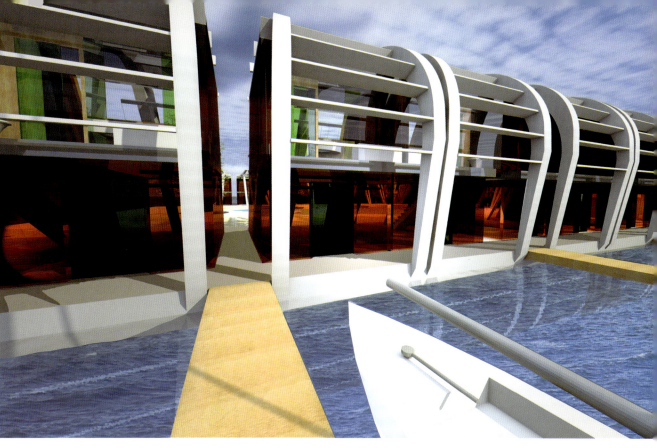

"We are tied to the ocean. And when we go back to the sea, whether it is to sail or to watch — we are going back from whence we came." –
John F. Kennedy, American president

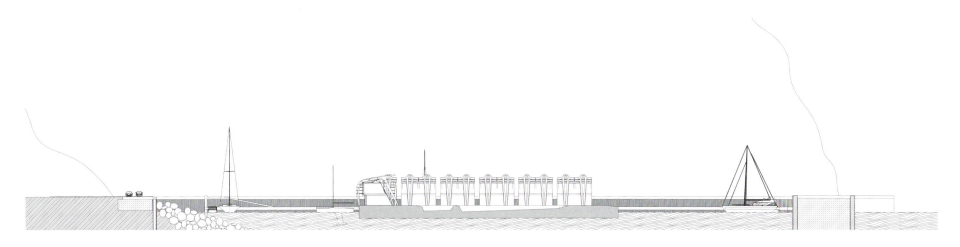

ONE OF ONE
HAMBURG, GERMANY

The shape of this houseboat is a direct result of the client's wish to incorporate maritime flair and develop an open accessible spatial arrangement. The interrelation of the two components house and boat, the inclusion of elements reminiscent of ship construction, and the large openings commonly found in house construction come together to form an independent coherent volume. A folded band gives the exterior its dynamic appearance and also creates an exciting play of open and closed surfaces. The floor plan is organized so that open-plan areas are sensibly separated from each other by carefully positioned stairways and service areas.

Architects: MONO Architekten, baubüro.eins
Structural engineers: Ingenieurbüro Holger Körner
Location: Uferstraße 8e, 22081 Hamburg, Germany
Completion: 2009
Client: Thorsten Freier
Gross floor area: 145 sqm
Function: office
Built on water: floats on dumb barge

Royal Dutch Shell PLC is currently building the world's largest floating facility. The bow and stern are half a kilometer apart — enough space for four football pitches.

Architects: büro13 architekten
Location: Marina, 17440 Kröslin, Germany
Completion: 2012
Client: Formstaal GmbH
Gross floor area: 104 sqm
Function: living
Built on water: floats on steel pontoon

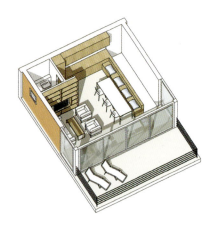

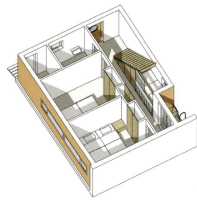

WATERKANTHUS 100
KRÖSLIN, GERMANY

Feel the gentle sway of water: These houses combine all the advantages of your own home: well-equipped, meticulously detailed, set in a marine surrounding or sea-landscapes. A floating structure, simply constructed, creating the foundation for a variable type of house that will fulfill all the requirements of contemporary living. The sliding glass frontage and floor-to-ceiling windows, wrapped around by a terrace, break the contrast of inner and outer worlds. The architects have created their first floating vacation homes in the Baltic Sea, in the harbor of Kröslin; a symbiosis of functionality and timeless design.

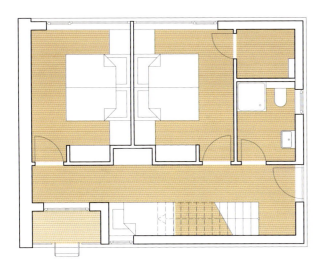

Dried totora reeds are used to make the floating islands on Lake Titicaca, Peru.

Architects: +31architects
Structural engineers: Dijkhuis Aannemersbedrijf B.V.
Location: De Omval 4, Amsterdam, The Netherlands
Completion: 2010
Client: confidential
Gross floor area: 197 sqm
Function: living
Built on water: floats on hollow concrete pontoon

DE OMVAL VILLA
AMSTERDAM, THE NETHERLANDS

This houseboat floats in the Amstel river in Amsterdam. The design manages to be contemporary without losing the characteristic appearance of the typical houseboat. Living on the water is becoming more and more popular in Holland. The clients wanted a boat with an open floor plan where they could enjoy views of the water and the outdoor space to a maximum. The distinguished curved line of the façade is controlled by regulations that state the boat can't be more than three meters above the water. The living area and open kitchen are located on the waterfront, from here one has a panoramic view of the Amstel. The split-level introduces an open route to the ground floor of the boat and allowed the creation of a terrace on the south side without exceeding the maximum building height.

"The sea, once it casts its spell, holds one in its net of wonder forever." – Jacques-Yves Cousteau, French oceanographer

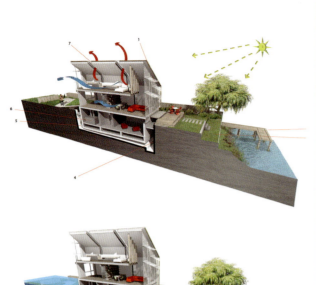

The UK's first Amphibious House is a pioneering, flood-resilient building, designed to respond to the uncertainties of future climate change. An amphibious house rests on fixed foundations, but can rise up to float in its dock if the site becomes flooded. A waterproof concrete hull creates a 'free-floating pontoon' upon which a lightweight timber frame house is constructed. The house is set between four steel guideposts. The low-energy dwelling has a south-facing façade overlooking the river and garden, which is formed from a series of stepped terraces that might flood incrementally to provide a flood warning system. Amphibious housing could be a solution to the uncertainty of future flooding on sensitive sites around the world.

AMPHIBIOUS HOUSE
MARLOW, UNITED KINGDOM

Architects: Baca Architects
Structural engineers: Techniker
Completion: ongoing
Client: confidential
Gross floor area: 225 sqm
Function: living
Built on water: supported by concrete piles and ground slab, with waterproof concrete hull

"If water were somewhat more rarefied, it could no longer sustain those prodigious floating buildings, called ships." – François Fénelon, French archbishop

Achitects:: C.F. Møller Architects
Completion: ongoing
Client: Aquadomi / H2Orizon
Gross floor area: 4,000–9,000 sqm
Function: hotel
Built on water: floats on concrete pontoon

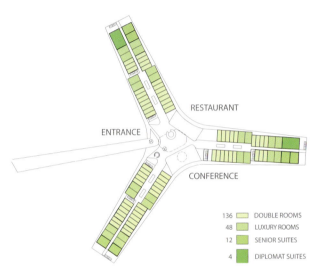

AQUADOMI HOTEL
VARIOUS

The AquaDomi Hotel series is based on a rational, prefabricated modular construction, built on floating concrete pontoons. As its name implies, the AquaDomi Star Hotel is star-shaped. The star shape secures views and balconies towards the water for all rooms, and can be rotated to suit different locations. A second, Z-shaped design version is adapted to smaller or narrower water areas, such as rivers and ports. The Z-shape allows similar advantages to the star shape, giving symmetrical balance and securing oblique views to the water, by avoiding an unattractive parallel alignment to the shore.

"Though inland far we be, our souls have sight of that immortal sea, which brought us hither." — William Wordsworth, British poet

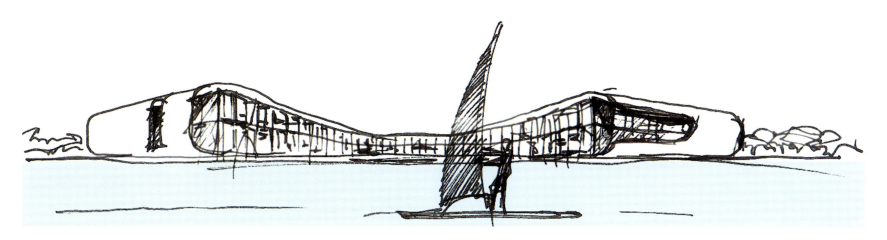

This is Germany's biggest floating building, situated in Zollhafen, Hamburg. The building moves with the tide, rising and falling three and a half meters and even floats with the water in a storm tide. The building rests on a concrete pontoon and the entrance is reached via a bridge. It offers a varied exhibition space spread over three levels. The three-story structures are made of steel and can be de- and re-constructed, this allows the building to be moved if necessary. The module frames are prefabricated and were assembled on the pontoon in just two weeks. The building requires no additional energy source. It functions entirely from the water of the river Elbe and from solar energy.

IBA DOCK
HAMBURG, GERMANY

Architect: Han Slawik Architekt / architech
Structural engineers: ims Ingenieurgesellschaft
Detail design: bof architekten
Energy concept: Immosolar
Location: Am Zollhafen 12, 20539 Hamburg, Germany
Completion: 2010
Client: Internationale Bauausstellung, IBA Hamburg GmbH
Gross floor area: 1,925 sqm
Function: exhibition and office
Built on water: floats on concrete pontoon

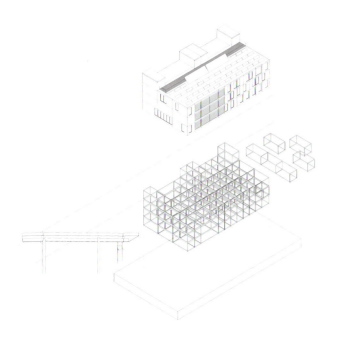
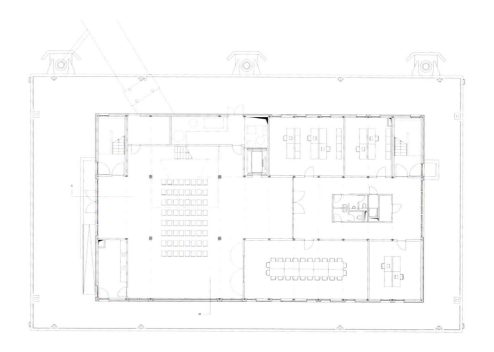

"Water is the driver of nature" –
Leonardo da Vinci, Italian Homo Universalis

The clients of this two-story houseboat wanted one big living space at the waterfront. The kitchen, living area and floating terrace are all located at water level. The floating terrace is an integral part of the living space, rigidly connected to the houseboat, and has the same floor finishing as the living space. The boat has a façade of taut aluminum cladding, where perforated panels on the quayside display the house number. LED lighting behind the perforated panels illuminates the façade in the evening and casts reflections on the water.

WATERVILLA WEESPERZIJDE
AMSTERDAM, THE NETHERLANDS

Architects: +31architects
Location: Weesperzijde, Amsterdam, The Netherlands
Completion: 2014
Client: confidential
Gross floor area: 200 sqm
Function: living

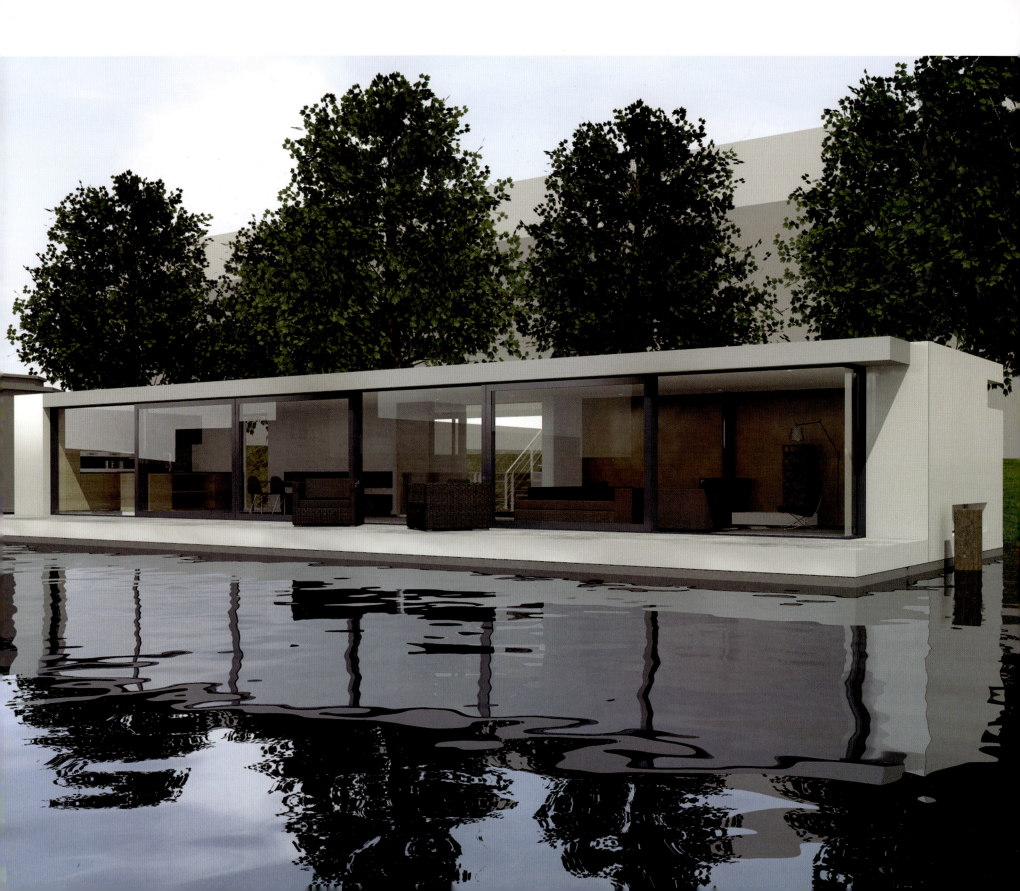

Flotation therapy is a form of therapy that is undertaken by floating in a warm salt water in a float tank.

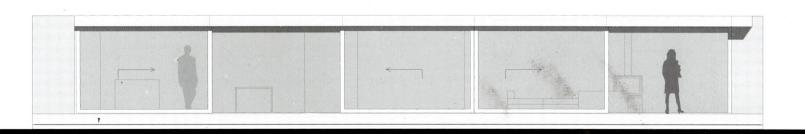

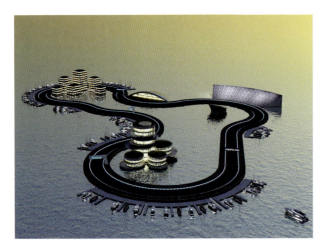

The floating racetrack is a revolutionary new concept for motor sport. This highly innovative design permits the track and its accompanying team and spectator facilities, including an eco leisure resort, to be re-configured into different layouts. The ability to dismantle the structures also allows the track to be leased and transported to different countries around the world with suitable international waterfronts. A waterproof concrete and expanded polystyrene pontoon system with steel bracing provides a stable base for the track, pits and grandstand. A floating marina allows for a substantial number of moorings and supporting luxury hospitality.

FLOATING RACETRACK
VARIOUS

Architects: Baca Architects
Completion: ongoing
Client: confidential
Function: race track
Built on water: floats on waterproof pontoon of expanded polystyrene

"It doesn't matter how long a tree trunk lies in the water, it won't become a crocodile." – African proverb

PRINSENSTEIGER
AMSTERDAM, THE NETHERLANDS

The flour pier, in front of the old flour factory at Amsterdam Westerdok (river IJ), was a forgotten piece of industrial heritage, when the idea was initiated to transform it to a public site. The flour pier was left in its original shape and designed to be open to the public from the boulevard as well as from a bridge joining the far ends – enabling visitors to stroll in a full circle. Topping the pier, a double storied volume was created – consisting of three luxury apartments. Glass façades and slightly shifted bay windows grant equal sunlight and spectacular views over the waterfront. The color white strengthens the desire for light and transparency and subtly hints to the core product of its past: flour.

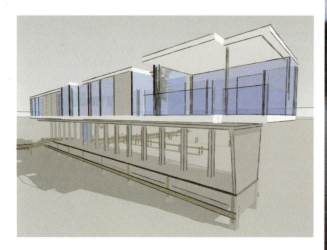

Architects: oth_architecten
Planning partner: Open Development
Initiative and design: Trude Hooykaas
Location: Kop Westerdok, Amsterdam, The Netherlands
Completion: ongoing
Client: confidential
Gross floor area: 500 sqm
Function: apartments and recreation
Built on water: supported by an industrial pier

Plankton is a collective term for organisms that drift or float in sea currents.

THE FLOATING EXPERIENCE
HAMBURG, GERMANY

Following the great success of the Floating Home realized in the City Sporthafen in Hamburg, the city has been enhanced by a floating conference center. The one-story structure contains two conference rooms. The two-story section houses a bar and lounge, as well as a multifunctional event room, kitchen and sanitary facilities. On the side facing the canal, the outer shell of the building is distinguished by generous window elements, which allow an unobstructed view of the water. On the shore side, the building ends with the rounded stern made of the white-coated façade panels, which are so characteristic of the Floating Homes. The spacious patios provide conference guests and visitors with an attractive and highly flexible outdoor area.

Architects: AMF Architekten Martin Förster
Location: Amsinckstraße 53, 20097 Hamburg, Germany
Completion: 2008
Client: Accor Hospitality Germany GmbH
Gross floor area: 1,010 sqm
Function: conference, bar and lounge
Built on water: floats on reinforced concrete pontoon

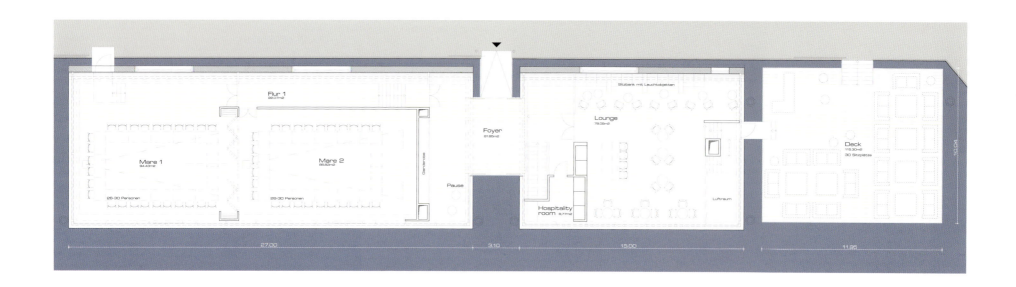

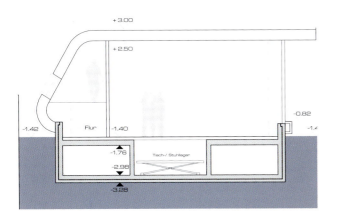

"While the river of life glides along smoothly, it remains the same river; only the landscape on either bank seems to change." –
Max Müller, German philologist and orientalist

DRIJF IN LELYSTAD
LELYSTAD, THE NETHERLANDS

Drijf in Lelystad consists of eight floating dwellings, for eight families. Having lived on water in their childhood, these families always dreamt of living on water once more. They united in a collective partnership called Float in Lelystad and commissioned eight different floating homes. All these families had their specific requirements. So all dwellings have their own characteristic size, color and shape. Direct contact with the water was a key focus of each design, along with: unobstructed views, split levels, abundant daylight, reflections of the water on walls and ceilings and water terraces on different levels. Color accents make one dwelling stand out from the others, giving the homes their own individual character.

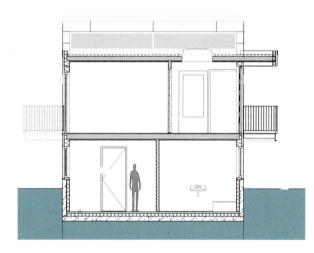

Architects: Attika Architekten
Structural engineers: ABC Arkenbouw
Location: Oeverzegge 1–15, Lelystad, The Netherlands
Completion: 2012
Client: Collective Partnership Drijf in Lelystad
Gross floor area: 195–285 sqm
Function: living
Built on water: floats on concrete cassion

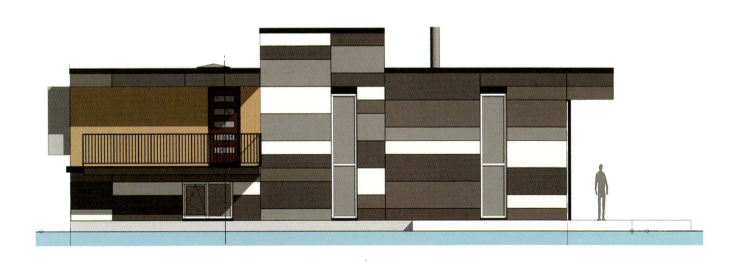

"A lake is the landscape's most beautiful and expressive feature. It is earth's eye; looking into which the beholder measures the depth of his own nature." – Henry David Thoreau, American writer

Restaurant "Het Bosch" has a superb location on the shore of De Nieuwe Meer lake, a stones throw away from Amsterdam's Zuid business district. The design has two distinct faces: one towards the marina and one facing De Nieuwe Meer. Both façades derive their qualities and appearance from the surroundings. The characteristic roof changes from three-pitched roof to a straight edge. This space saving configuration makes it possible to fit the desired program into the tight prescribed envelope. The building itself was constructed almost completely out of pre-fabricated wooden elements. By using this durable and prefab construction, the building time is significantly reduced, important for the restaurant's operational results.

HET BOSCH
AMSTERDAM, THE NETHERLANDS

Architects: Dreissen architects, JagerJanssen architects
Structural engineers: Lüning
Location: Jollenpad 10, 1081 KC, Amsterdam, The Netherlands
Completion: 2010
Client: Het Bosch
Gross floor area: 480 sqm
Function: restaurant, boathouse, penthouse
Built on water: supported by steel columns

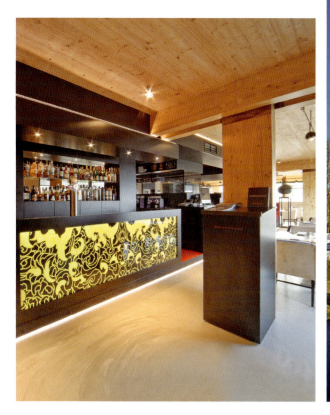

Almost a third of the Netherlands is situated below sea level.

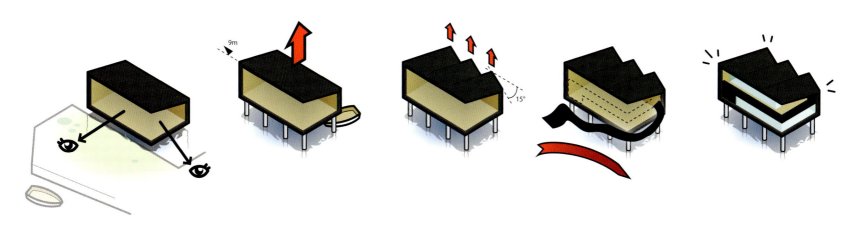

Architects: Baca Architects, Waterstudio.NL
Structural engineers: WSP and Ramboll
Location: Royal Victoria Dock, London, United Kingdom
Completion: ongoing
Client: Hadley Mace
Function: mixed use
Built on water: floats on concrete pontoons

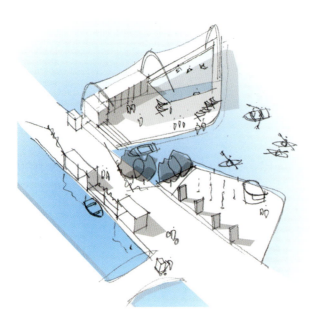

THE CROWN, FLOATING VILLAGE
LONDON, UNITED KINGDOM

London's floating village is conceived as a 'Crown' in the Royal Docks, London. Part of a scheme to reinvigorate this East London waterspace, the village not only creates valuable new residential and commercial space in the busy capital, but introduces a new approach to water-based design. Evolving the model of the traditional English village, water is the development's backbone. The entire settlement will be almost self-sufficient, just as a village is an island within the countryside, but planned along canals, rather than roads. The development is supported by a comprehensive "waterspace plan", which provides the framework for sustainable redevelopment of the docks and establishes the principles floating cities around the world.

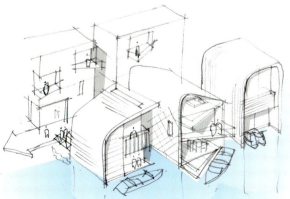

"Water is the principle of all things; and God is that mind which shaped and created all things from water." — Cicero, Roman philosopher

Architects: 5468796 Architecture
Structural engineers: Wolfrom Engineering Ltd
Location: Storm Bay, Ontario, Canada
Completion: 2011
Client: confidential
Gross floor area: 279 sqm
Function: boatport and raised deck

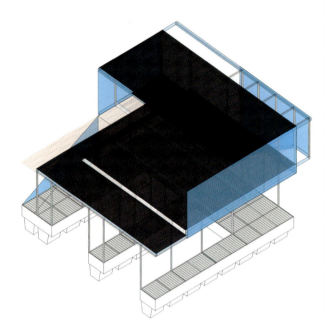

GUERTIN BOATPORT
ONTARIO, CANADA

The Guertin Boatport is a two-story, open-air floating dock and fixed boardwalk located in Storm Bay, western Ontario. The main level provides two sheltered boat stalls while the upper floor serves as an informal lounge space and viewing deck. The project is composed of fragmented vertical planes clad in reflective materials that scatter and redirect the light. Perforated metal walls protect against the wind while still allowing for extended views of the bay. The boatport is accessed by an extruded aluminum plank boardwalk that follows the profile of the rocky shore. Flexible, hinged joints accommodate the rise and fall of water levels, as well as the freeze and thaw cycles of ice.

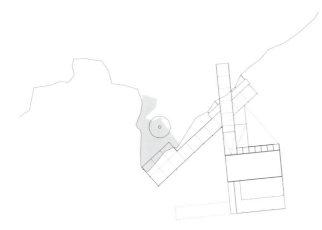

"In rivers, the water that you touch is the last of what has passed and the first of that which comes." – Leonardo da Vinci, Italian Homo Universalis

Floating Eco House is a mobile house with minimal environmental impact based on a modular structure. The project uses low-energy solutions and environment-friendly materials. Its construction makes it possible to change the appearance and building size at any time during its lifecycle. The house can also be rearranged to meet the needs of its residents. The possibility of separating different materials allows a trouble-free disposal, recycling and upcycling. Floating Eco House is designed to meet the criteria of an energy-autonomous building. This means that the house can function independently of urban technical infrastructure and uses off the grid technology.

FLOATING ECO HOUSE
ANYWHERE

Architects: Monika Wierzba
Completion: 2011
Client: confidential
Gross floor area: 14 sqm / module
Function: living
Built on water: floats on hydro concrete pontoon

"Return to old watering holes for more than water; friends and dreams are there to meet you." – African proverb

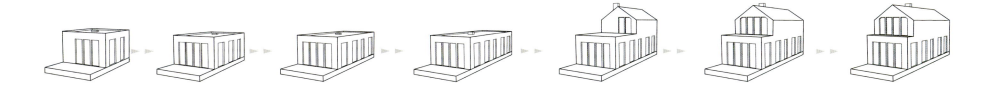

Architects: Erkko Aarti, Arto Ollila & Mikki Ristola (AOR)
Structural engineers: Price & Myers
Location: Camley Street Natural Park, London, United Kingdom
Completion: 2014
Client: Camley Street Natural Park
Gross floor area: 30 sqm
Function: culture
Built on water: floats on custom-made steel pontoon

VIEWPOINT
LONDON, UNITED KINGDOM

Viewpoint attempts to unite nature and architecture on London's Regent Canal. An island of calm amidst the hustle and bustle of London's King's Cross, Viewpoint acts as a viewing platform and learning facility for Camley Street Natural Park. AOR architects Erkko Aarti, Arto Ollila and Mikki Ristola based the structure on traditional Finnish Laavus, which are shelters used during hunting and fishing trips. It comprises a small cluster of triangular volumes that form hideaways and seating areas. The outer surfaces are clad with rusty corten steel, creating a dialogue with canal boas seen in the area. Inside, triangular peepholes at eye level offer glimpses of birds such as swans and kingfishers.

Jesus bugs are able to 'walk' on water due to the presence of fine hairs on their legs.

RADIO YEREVAN HOUSEBOAT ON THE EILBEK CANAL
HAMBURG, GERMANY

This location on the Eilbek Canal has a rather introverted character. For this design, the architects deliberately steered clear of dynamic design elements that suggest movement and tried to use materials and construction techniques to create a link to the subject of maritime. The main design concept is the idea of a shipwreck, the rump of which is corroding in the water and slowly disintegrating. The structure supporting the lower level is mounted onto a concrete pontoon. The patina on the corten steel cladding changes depending on the weather, time of day or season. The private rooms are located on the lower level, while the living and bedrooms are oriented towards the water. A large wooden terrace can be found on both sides of the glass box.

Architects: sprenger von der lippe Architekten
Location: Uferstraße 2c, 22081 Hamburg, Germany
Completion: 2010
Client: Wiebke Toppel / Haig Balian
Gross floor area: 180 sqm
Function: living
Built on water: floats on concrete pontoon

"The mind is like an iceberg; it floats with one-seventh of its bulk above water." –
Sigmund Freud, Austrian neurologist

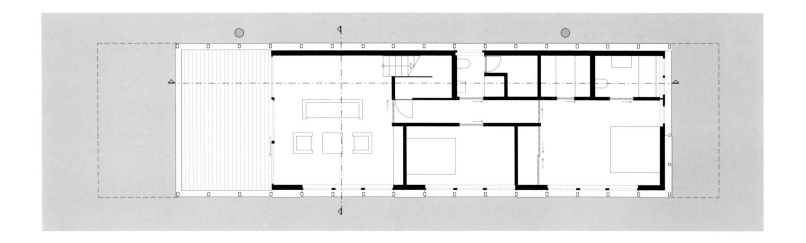

Architects: Baumhauer Eichler Architekten
Location: Humboldtinsel, 13507 Berlin-Tegel, Germany
Completion: 2015
Client: Martrade Immobilien, Dusseldorf
Gross floor area: 266 sqm / house
Function: living
Built on water: floats on concrete pontoon

FLOATING HOUSES HUMBOLDTINSEL
BERLIN, GERMANY

Cities are currently experiencing a renaissance as attractive locations for new ways of living, working and spending free time. Brownfields are being transformed into scientific campuses, decommissioned railway lines to active parks, and old docks into residential areas for "living on water". As early as 1984, such a project was developed in the Post-Modern style in Tegeler Harbor in Berlin for the International Architecture Exhibition (IBA). The new Humboldtinsel project is a new modern ensemble of floating houses, pier houses and water apartments that stretches over 600 meters along the former quay wall. They float on concrete pontoons and are connected to the island's district heating network.

"The wise find pleasure in water." –
Confucius, Chinese philosopher

This office building floats amidst the Waternet fleet of vessels that clean the waters of the Amsterdam canals each day. The office program has been realized on two connected floating concrete caissons. In total the building measures 31 by 12 meters and has three floors, making it the biggest ark in the Netherlands. It houses office space on the first floor and showers and locker rooms in the underwater basement. A spacious double height canteen forms the heart of the building and links all rooms on ground and first floor level with the harbor.

FLOATING OFFICE FOR WATERNET
AMSTERDAM, THE NETHERLANDS

Architects: Attika Architekten
Structural engineers: ABC arkenbouw
Location: Papaverweg 54, Amsterdam, The Netherlands
Completion: 2011
Client: Waternet
Gross floor area: 875 sqm
Function: office
Built on water: floats on concrete pontoon

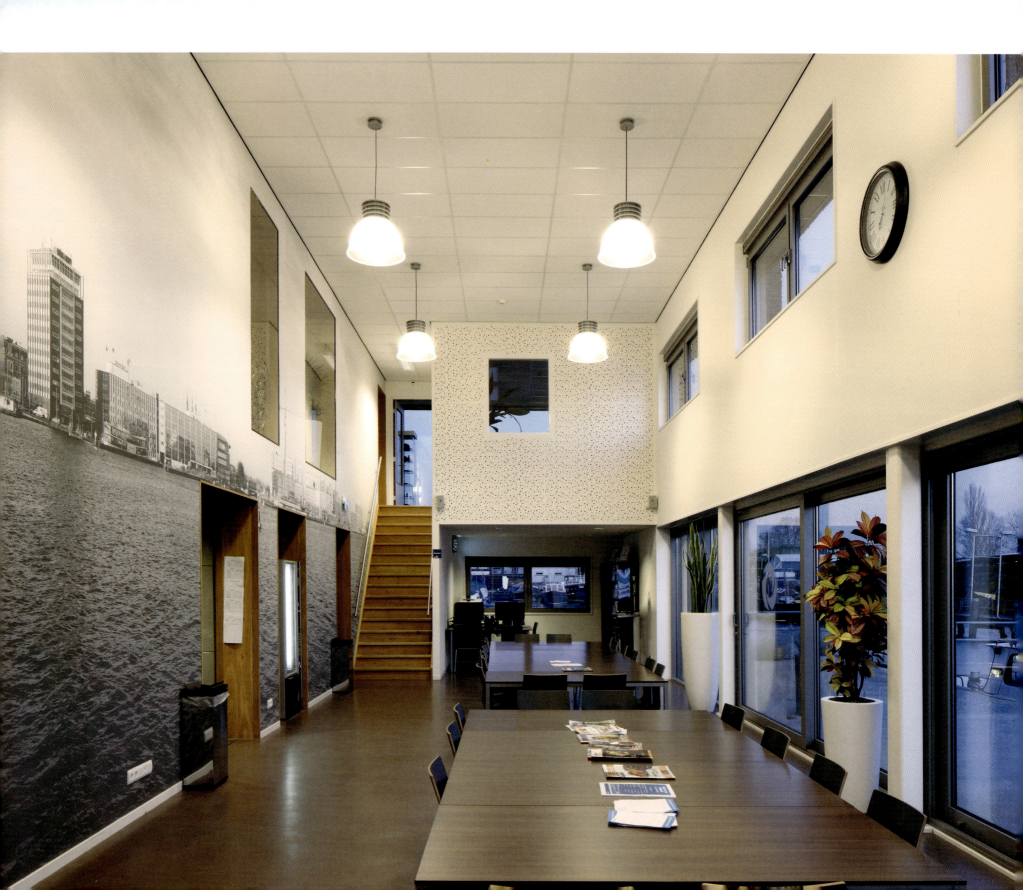

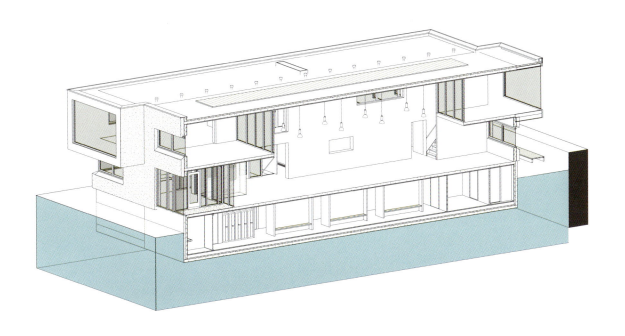

140 << 141

There are more atoms in a glass of water than glasses of water in all the oceans on Earth.

BROOKLYN BRIDGE PARK PIER 5
BROOKLYN, NY, USA

Reclaiming 34 hectares of Brooklyn's post-industrial waterfront, the Brooklyn Bridge Park continues the tradition of New York City's grand parks. Like a beacon, the pavilion on Pier 5 broadcasts its vital multi-use role and harbor location. The design is lightweight and is designed to reuse the existing pier pilings to support it. The floating site minimizes energy consumption by capitalizing on the site's natural resources. A heat exchange pump using the river water feeds the radiant cooling and heating floor system while large operable doors along the east and west walls harvest prevailing winds while further connecting the building to its extraordinary context and views of Lower Manhattan and the Brooklyn Bridge.

Designers: James Carpenter Design Associates
Structural engineers: Schlaich Bergermann und Partner
Landscape architects: Michael Van Valkenburgh Associates
Location: Joralemon St, Brooklyn, 11201 NY, USA
Completion: future
Client: Brooklyn Bridge Park Development Corporation
Gross floor area: 1,895 sqm
Function: recreation
Built on water: supported by piers

"Hovering at the water's surface surrounded and infused with a site's unique qualities of light, a floating structure's buoyancy intimately reconnects us with a fundamental experience of place defined by nature." –
James Carpenter, American artist and architect

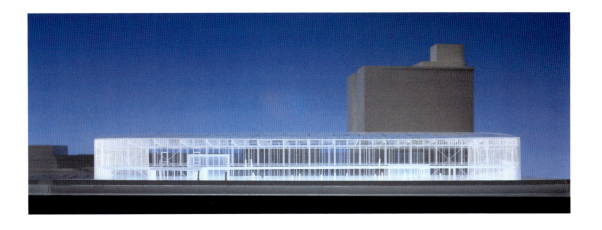

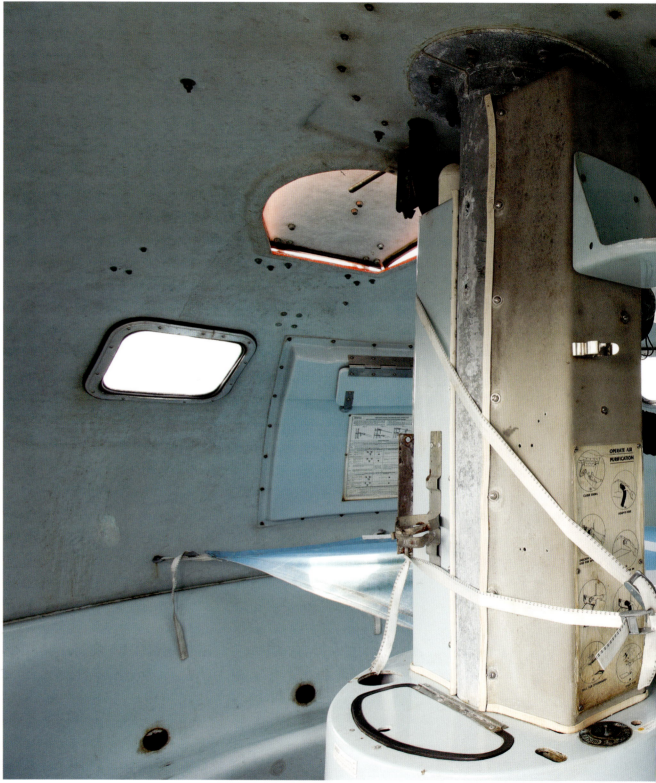

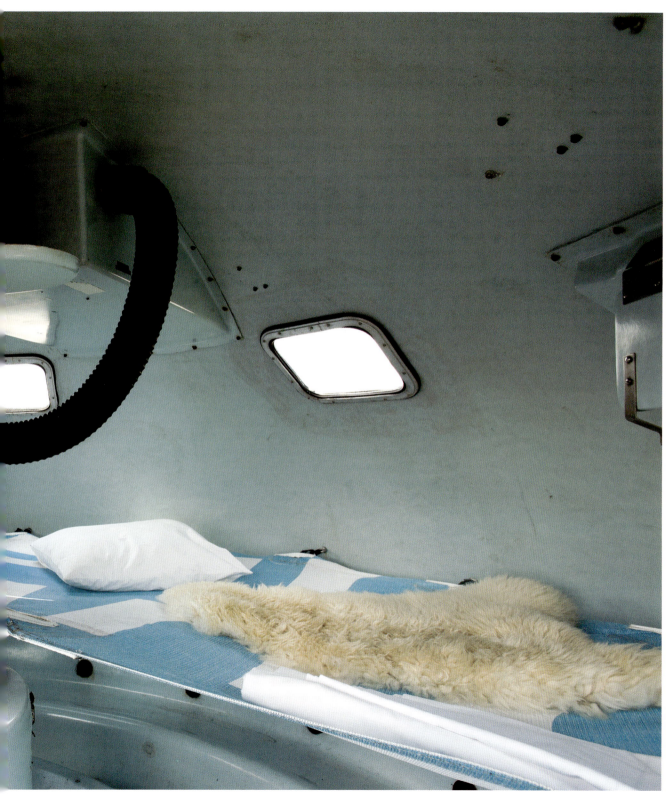

Not even a fixed address identifies the orange-colored objects that currently float in The Hague's harbor like UFOs after an emergency landing. Originally designed for 28 persons, the escape capsules can accommodate a maximum of four guests each and offer a choice between hammock or sleeping bag. The unique survival experience can be even more buoyant with the aid of champagne from the "luxury survival suitcase", which is available alongside a DVD player, a karaoke machine or a lit disco ball.

CAPSULE HOTEL
VARIOUS

Architects: Denis Oudendijk
Completion: 2005
Client: Denis Oudendijk
Gross floor area: 13 sqm
Function: hotel
Built on water: floating plastic capsules

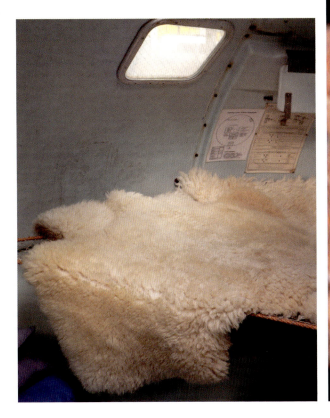

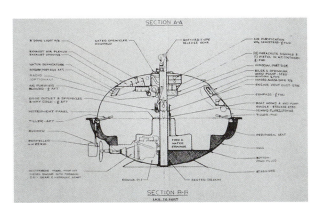

"Be like a rock in the middle of a river, let all of the water flow around and past you." – Zen saying

TAFONI FLOATING HOME
SAUSALITO, CA, USA

The primary goal of this conceptual project is to change attitudes towards living on a houseboat and promote a lifestyle that encourages the preservation of green fields. Tafoni is spacious, yet compact. Typical houseboats have low ceilings and often feel cramped, which can detract from the comfort many residents desire of their homes. Tafoni is a multi-purpose living pavilion that can serve as a permanent house, a weekend retreat, a relaxing summer destination or a place to entertain friends and hold business parties. This houseboat is located in Sausalito, California, an area that serves as a permanent residence for many and is an example of peaceful coexistence between humans and nature, appropriately regulated by local laws to ensure protection of the environment.

Designer: Joanna Borek-Clement
Design completion: utopia
Gross floor area: 89 sqm
Function: living

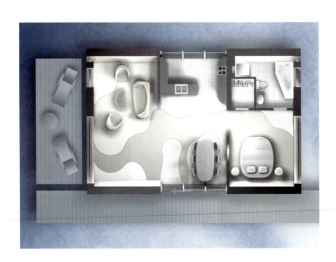

"Fish must swim." – Petronius, Roman writer

The free-floating ecolodge is an ecologically responsible accommodation with a minimal impact on the surrounding nature. It is a unique design, created by Marijn Beije Design. All wood used on the shell is Western Red Cedar with FSC certification. The safety glass is toughened and laminated. There is space for a boat to be docked against the long side. The lodge can float freely in the water or it can be connected to a pier. Ecolodge is self-sufficient, with power provided by solar panels. A toilet, kitchenette and a fixed seat with underwater views of the lake are all located in one volume, while the other houses four beds. The terrace between the two volumes is approximately 25 centimeters above water level. A ladder leads from the terrace to the 'crow's nest' above.

ECOLODGE
DORDRECHT, THE NETHERLANDS

Architects: Marijn Beije
Location: Biesbosch National Park, Dordrecht, The Netherlands
Completion: 2012
Client: confidential
Gross floor area: 58 sqm
Function: living
Built on water: floats on aluminum hull

"Ignorance is an ocean and knowledge is a raft that floats upon it." – Icelandic proverb

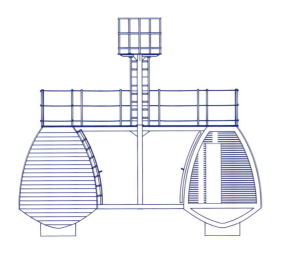
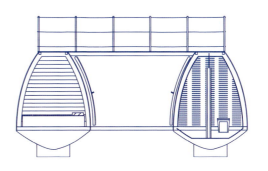

Pampus Harbor is a little used 'port of refuge'. It is a strategic spot located between Almere and Amsterdam, which in the future will transform into a waterfront colony within the visionary plans of the development of the Amsterdam-Almere area. This context and the character of the location are the basis for developing a specific living typology for people who are drawn to an environment with more freedom and independence. The study proposes 500 floating houses, which give a maximum of individuality and a sense of freedom created by the way the individual houses are positioned. Based on several floating techniques and organization principles, five typologies have been developed with a maximum relation to the water.

PAMPUS HAVEN
PAMPUS HAVEN, THE NETHERLANDS

Architects: MVRDV
Structural engineers: Mervac Maritime, Bergen
Completion: ongoing
Client: Het Oosten Kristal, Amsterdam and Municipality of Almere
Gross floor area: 4,500 sqm
Function: living
Built on water: supported by platform resting on riverbed

"In matters of style, swim with the current; in matters of principle, stand like a rock." –
Thomas Jefferson, American president

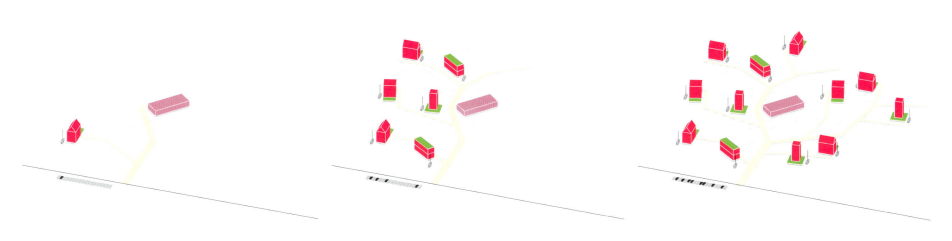

This humble teahouse stand in the lush ShiQiao Garden. The garden and Tea House embrace traditional elements of Chinese garden design, while also blending into the natural environment. The focal point of the design is the courtyard, which uses bamboo to create an interesting play of vertical and horizontal lines. Each of the spaces offers views of the surrounding lake. The use of natural materials such as bamboo and brick reduces energy requirements and has a low impact on the environment. The pocket of voids improves natural ventilation within the bamboo courtyard while the thick brick wall retains heat in winter, reducing the dependency of mechanical heating and cooling system. At night, the light shines out through the walls, giving the teahouse the appearance that it is floating on rather than standing in the lake.

TEA HOUSE – BAMBOO COURTYARD
YANGZHOU, CHINA

Architects: HWCD
Location: ShiQiao Garden Yangzhou, Jiangsu, China
Completion: 2012
Client: Construction Bureau of the Economy and Technology Development District, Yangzhou
Gross floor area: 400 sqm
Function: leisure

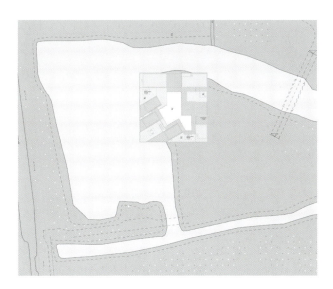

"Water too pure breeds no fish." –
Mao Zedong, Chairman of the Communist Party China

DOCKS - CITE DE LA MODE ET DU DESIGN
PARIS, FRANCE

The Docks of Paris is a long, thin building built in concrete at the turn of the last century. The city of Paris launched a competition to create a new cultural program and building on this site. Jakob+MacFarlane opted to retain the existing structure and use it to inspire and influence the new project. The concept of the new project was to create a new external skin that is inspired primarily by the flux of the Seine and the promenades along the sides of the river banks. This skin is created principally from a glass exterior skin, steel structure, wood decking and grassed, faceted roofscape. The design not only exploits the maximum building envelope but enables a continuous public path to move up through the building.

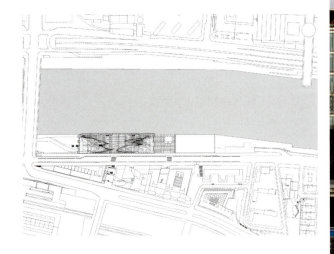

Architects: Jakob+MacFarlane
Location: Dock d'Austerlitz, Paris, France
Completion: 2012
Client: Icade G3A
Gross floor area: 20,000 sqm
Function: education, culture, shops
Built on water: supported by platform beneath

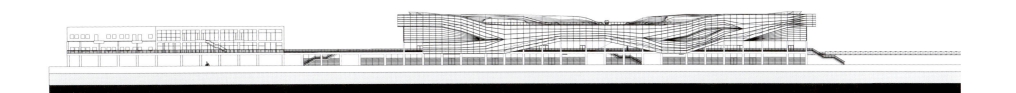

"Praise the sea; on shore remain." –
John Florio, British linguist and lexicographer

FLOATING CINEMA
LONDON, UNITED KINGDOM

In June 2012, UP Projects launched an open international design competition, in partnership with the Architecture Foundation, to design a new Floating Cinema for London's waterways that would launch during summer 2013. Duggan Morris Architects won the competition with the proposed design entitled "A Strange Cargo of Extra-Ordinary Objects". The design was not only practical, but also a highly inventive and imaginative response to the brief. The architects proposed that the floating cinema advocates films as a 'precious cargo' that might connect communities and utilize the waterways to encourage discourse between people and places through the Lea Valley and beyond.

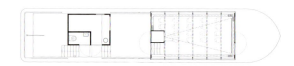

Architects: Duggan Morris Architects
Structural engineers: Price & Myers
Completion: 2013
Client: UP Projects, supported by The Legacy List with corporate partner Bloomberg
Gross floor area: 21 sqm
Function: entertainment
Built on water: floats on steel pontoon

The Philippines are made up of around 7,200 islands.

THEMATIC PAVILION
YEOSU, SOUTH KOREA

The Expo 2012 Yeosu eyeOcean pavilion was conceived as a floating body, quietly resting on the waves. The body is stabilized by programmable water jets securing the exact position and balance of the pavilion. The public paths follow the internal soap bubble structure, shaped like miniature valleys between the upward curving membranes of the internal cell structure and are highlighted for orientation during the night. The dark skinned body is kept wet by the cyclic eruptions of the exhalation opening in the central Iris, making it shine and reflect the sunlight. Water drips continuously from the edges to create a cooling effect on the dark surface.

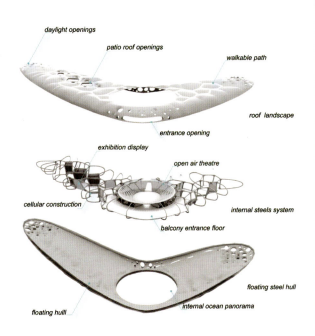

Architects: ONL [Oosterhuis_Lénárd]
Location: Expo 2012 Yeosu, South Korea
Completion: 2009
Client: confidential
Gross floor area: 6,180 sqm
Function: cultural
Built on water: supported from underneath and floats on wooden platform

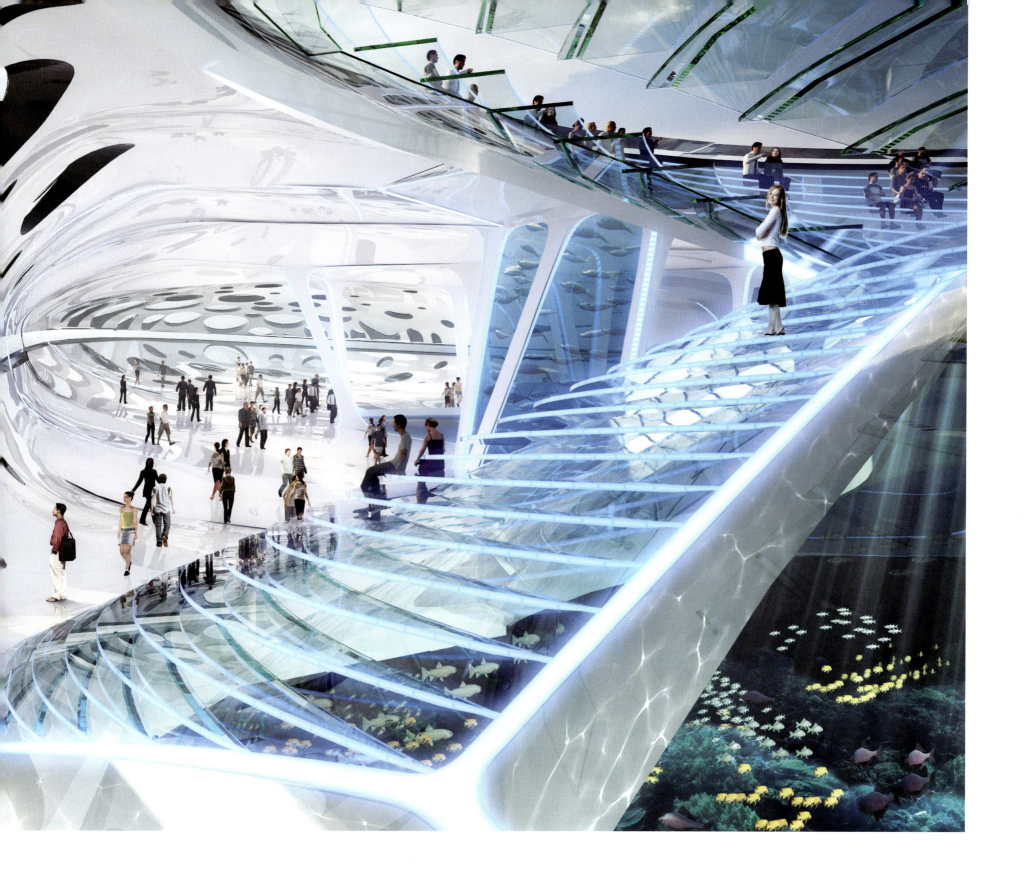

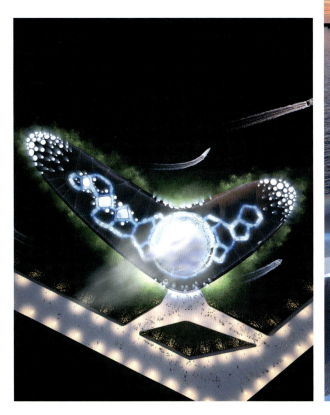

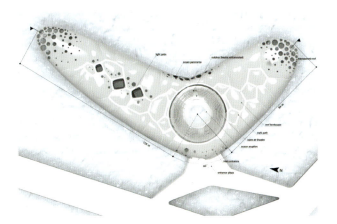

The Dead Sea is located 416 meters below sea level.

Architects: Robert Nebolon Architects
Structural engineers: Sarmiento Engineering Inc.
Location: 300 Channel Street, San Francisco, CA 94158, USA
Completion: 2013
Client: confidential
Gross floor area: 193 sqm
Function: living
Built on water: floats on concrete hull

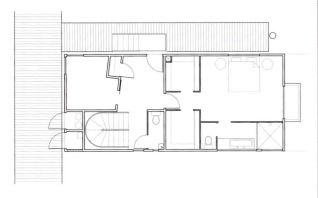

FLOATING HOUSE
SAN FRANCISO, CA, USA

The San Francisco Floating House is located in Mission Creek, an obscure backwater canal within the San Francisco city limits, once surrounded by acres of tidal shoals, and later, industrial zones. Nested permanently in a protective, stabilizing concrete barge, the metal-sided house is like a New York City loft on the water. Recalling the area's industrial past, the house has modern metal paneling and a saw-tooth roof; features commonly found in large older factories nearby. Inside the white volume, one large room can be found with the social areas where friends and family all gather. Large warehouse windows allow views of the water below and the San Francisco Skyline on the other side of Mission Creek.

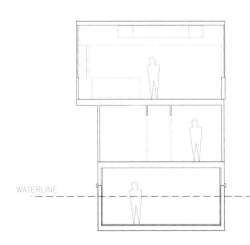
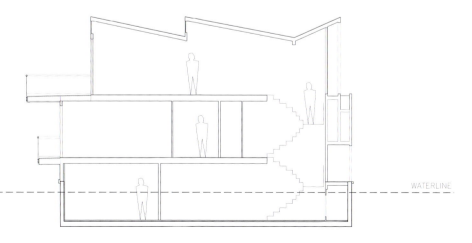

"When you freefall for 7,000 feet it doesn't feel like you're falling: it feels like you're floating, a bit like scuba diving." – Benedict Cumberbatch, British actor

Architects: Nio Architecten
Structural engineers: Waterhuys
Location: Jaagpad 17, Amsterdam, The Netherlands
Completion: 2012
Client: confidential
Gross floor area: 240 sqm
Function: living, music studio
Built on water: rests on concrete supports

TWIN BLADE
AMSTERDAM, THE NETHERLANDS

Seen from the outside, this house appears to be almost symmetrical. But once inside, your perception capsizes to reveal an asymmetrical structure with two faces. This houseboat has been designed for a couple with completely different characters; one is a visual artist, the other a composer and musician. One studio is located below ground level, while the other is on the upper level. This allows the couple to share the space and focus on their work without getting in each other's way. The staircase is a key element of the design, joining the various different areas.

Archimedes realized that a floating object displaces its own weight in a fluid. As he stepped into the bathtub, water splashed over the side. The weight of the water equaled his body weight. This idea is known as Archimedes' principle.

HAUSBOOT ON THE EILBEK CANAL
HAMBURG, GERMANY

The spatial design of this houseboat is determined by the special characteristics of the location. The main level lies just above the water line and presents a closed face to the canal bank, but opens out on the other three sides to make the most of views across the water. The upper floor opens out onto a large terrace, successfully merging inside and outdoor spaces. The house has a rather meandering spatial arrangement, with wooden panels used to form the floors, walls and ceilings. The rooms are arranged to make the most of the views across the water.

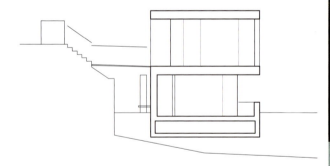

Architects: martinoff architekten
Location: Uferstraße 2b, 22081 Hamburg, Germany
Completion: 2010
Client: confidential
Gross floor area: 180 sqm
Function: living
Built on water: floats on reinforced concrete pontoon

The "Friendly Floaters" are a host of rubber ducks and other platic animals that became famous after they were washed overboard during a storm and have been floating in the ocean ever since.

Architects: Nio Architecten
Completion: 2005
Client: Maarten Aris, Amsterdam
Gross floor area: 80 sqm
Function: living

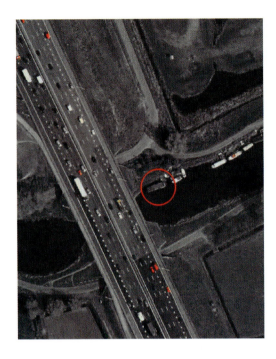

POINT ZERO
AMSTERDAM, THE NETHERLANDS

This houseboat occupies a niche in the Dutch housing market. Architects always nurture hope that the people who are looking for different types of housing – in empty offices, obsolete industrial buildings (lofts), old farms, holiday homes, caravans and houseboats – will gain the upper hand and that the typical housing format, which is fixed in the brains of real estate agents, will start to melt. But it is most likely that this houseboat is once again an example that the gap between the desire for specific individual living and the serial development of house building with the related rules is too big to fill.

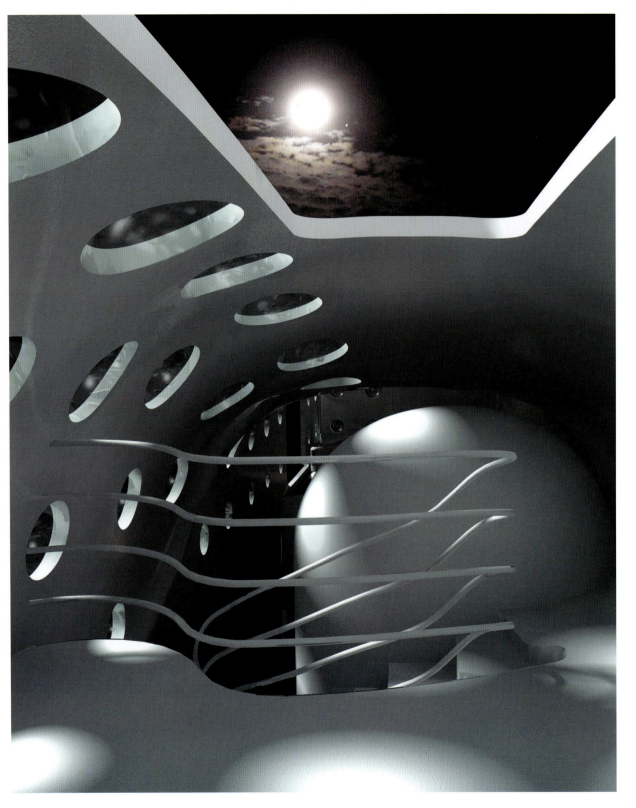

Ships float higher in sea water than in fresh water because salt makes the sea water more dense.

BALTIC SEA ART PARK
PÄRNU, ESTONIA

The urban role of the Baltic Sea Art Gallery is to be a new, visible sign in a panorama of the city in a form of a gate, which emphasizes the connection between the city center and the river. The idea was to create a floating square, which forms the end of the path leading from the Old Town towards the riverside. The floating piazza is designed as a public space in the form of a market square that completes the urban structure of Pärnu. It is a multifunctional space that serves variety of uses – a stage for different kinds of cultural events or just a meeting point for city dwellers.

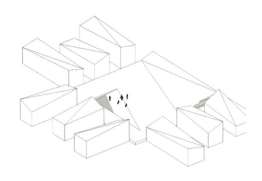

Architects: WXCA
Location: Lai Street 2-8, Pärnu, Estonia
Completion: ongoing
Client: Museum of New Art
Gross floor area: 1,295 sqm
Function: exhibition
Built on water: floats on concrete pontoon

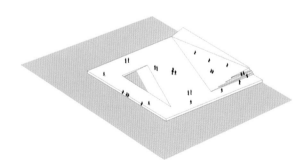

Approximately 70% of the earth's surface is covered with water.

Architects: Rost.Niderehe Architekten I Ingenieure
Static timber construction, balance calculation: Stephan Niderehe
Static pontoon, ship component verifcation: Buschmann und Söhne
Location: Uferstraße 8c, 22081 Hamburg, Germany
Completion: 2010
Client: Amelie Rost and Jörg Niderehe
Gross floor area: 130 sqm
Function: living
Built on water: floats on steel hull

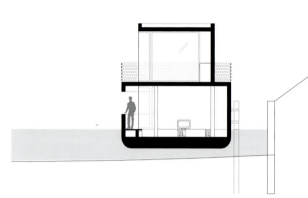

HOUSEBOAT ON THE EILBEK CANAL
HAMBURG, GERMANY

The idea behind this design was to create a house that shared some fundamental similarities with the character of a boat. The clear, simple design is reflected in the floor plan and the spatial organization. A single wall aids the arrangement of the rooms from public to private, from outside to inside, and from above to below. This wall spirals from the outer walls to the center of the house and encloses the bathrooms, private areas and laundry room. A large roof terrace, living and dining areas are all located on the upper floor, while the lower level houses the private areas, such as a study, bedrooms and living room. Wood and steel were chosen as the main materials, as these are also used in traditional ship construction.

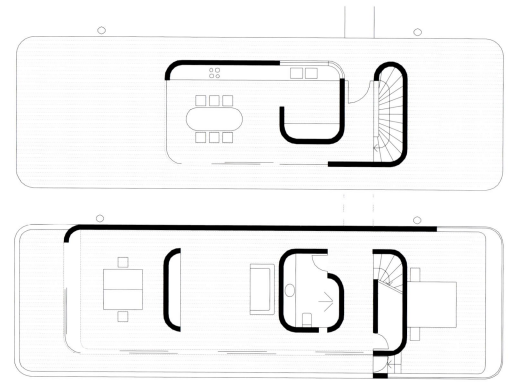

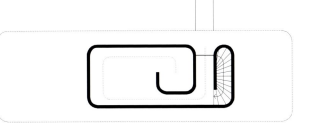

Lemons float, limes sink.

RE:Villa is a proposal for a floating habitat constructed from waste materials. The construction principle is based on yacht building techniques and draws together building engineering and maritime architecture. With the success of the prototype more sustainable and flood-proof habitats can be realized. The design for the prototype is a luxury floating villa. Wide panoramic views show the beauty of the coastline. The new landscape is 1,260 square meters and allows inhabitants to grow their own food and to enjoy the outdoors. Beside seaweed cultivation, recycling, solar- and wave energy, agriculture is an important factor that helps make the floating villa self-sufficient.

RE:VILLA
VARIOUS

Architects: WHIM Architecture
Completion: utopia
Gross floor area: 1,260 sqm
Function: living
Built on water: floats on hollow chambers

"It is not that life ashore is distasteful to me. But life at sea is better." –
Sir Francis Drake, English admiral

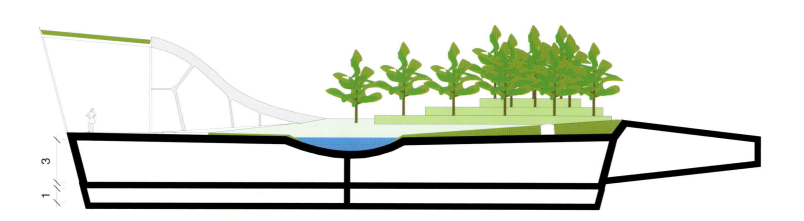

Architects: SLO Architecture
Completion: 2013
Client: NYC Department of Parks and Recreation /
Gowanus Canal Conservancy
Gross floor area: 40 sqm
Function: exhibition
Built on water: floats on 128 two-liter plastic soda bottles

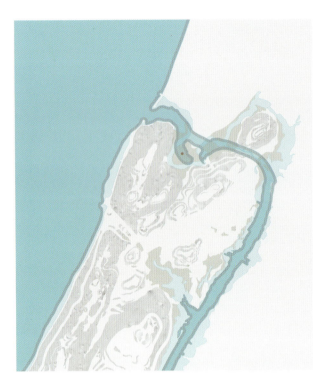

HARVEST DOME 2.0
NEW YORK CITY, NY, USA

Harvest Dome is a giant diaphanous orb for the New York City waterways. A cupola constructed from over 450 umbrellas frames, and made buoyant with a ring of 128 empty two-liter soda bottles, Harvest Dome transforms the eight-pointed steel frame into the transcendent form of an architectural dome for the water, to float alongside and bring attention to the tidal salt marshes of the city. Harvest Dome 2.0 was constructed during July of 2013 at the Brooklyn Navy Yards. On July 31, 2013, the dome was tugged up the East and Harlem Rivers via barge and then guided into the center of the inlet with the help of the Inwood Canoe Club.

"Not every lake dreams to be an ocean. Blessed are the ones who are happy with whom they are." – Mehmet Murat ildan, Turkish playwright

Private islands are the unique embodiment of luxury and privacy and offer a freedom not found in any other kind of property. However, there is also a downside to private islands on the market, because many of them are too remote and the construction and maintenance of the facilities is expensive. Dutch Docklands is changing the traditional private island market by creating Floating Private Islands: any shape, any place. Working in cooperation with Christie's International Real Estate, they are the only company in the world offering private floating islands, tailor-made to suit the client's specifications.

AMILLARAH – FLOATING PRIVATE ISLANDS
MALDIVES; DUBAI, UAE; MIAMI, USA

Architects: Waterstudio.NL
Developers: Dutch Docklands International
Completion: future
Client: various
Function: living
Built on water: floats on concrete foundation

The swim bladder is an organ found in teleost fishes. It serves to balance out the specific gravity of the fish in the surrounding water.

SEOUL FLOATING ISLANDS
SEOUL, SOUTH KOREA

The Seoul Floating Islands are intended to revitalize the Han River that divides Seoul, uniting the northern and southern halves of the city by activating the seam between them. The design of the islands is inspired by the phases of a flower: a seed, a bud, and a blossom. Each of the islands takes on the form of one of these stages as delicate structures of glass and steel. The islands and steel frames were erected on the riverbank and then rolled into the river where they are secured by an intricate mooring and pontoon system. The Seoul Floating Islands are a landmark and key amenity for Seoul, attracting world class events as well as hosting everyday local activities.

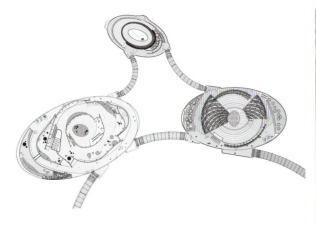

Architects: Haeahn Architecture + H Architecture
Structural engineers: Opus Pearl
Location: Han River, Seoul, South Korea
Completion: 2011
Client: Seoul Metropolitan Government
Gross floor area: 9, 210 sqm
Function: culture
Built on water: floats on pontoon system

Ernie's Rubber Duckie first appeared in Sesame Street in 1969.

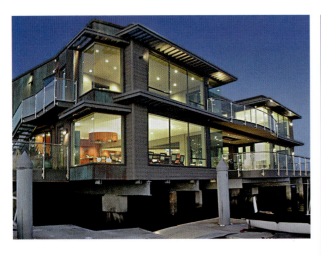

This elegant modern home captures the magic of living in the San Francisco Bay Area. From the street, the house offers no clue to visitors as to the spectacular views over the bay. Constructed on a pier over the water, the home gently suggests ship's hulls and seashells. Once inside, the entire house dissolves into glass, framing views of downtown San Francisco, the Golden Gate Bridge, and the water. A double-height living room and mezzanine open up through stacked sliding glass doors to an expansive deck and balcony for integrated indoor and outdoor living. Corrugated copper siding with verdi-gris was chosen for its durability to salty marine conditions. The cedar wood siding was specified to be rough-surfaced to absorb more wood-stain.

SAN FRANCISCO BAY HOUSE
RICHMOND, CA, USA

Architects: Robert Nebolon Architects
Structural engineers: Young&Burton Contractors
Location: 1456 Sand Piper Spit, Point Richmond, CA, USA
Completion: 2006
Client: confidential
Gross floor area: 648 sqm
Function: living
Built on water: supported by concrete piers

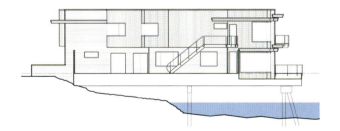

"Water is the best of all things." –
Pindar, Ancient Greek lyricist

AALBORG HAVNEBAD
AALBORG. DENMARK

Aalborg Havnebad forms part of Aalborg's recreational waterfront with its parks, squares, sports and play areas, exhibition buildings, leisure and entertainment buildings. The harbor bath is equipped with a children's pool for the smallest children, a play pool for somewhat older children, and a 60-meter-long pool for adults, which flows into a diving pool and a winter pool. The harbor bath contains winter bathing facilities, including a winter pool, a building for lifeguards, changing rooms, a clubroom and sauna. The winter pool is surrounded by an arcade, on top of which a balustrade provides a view of the harbor, the city and the baths. The design comprises floating pontoons secured by a low structure of telescopic poles.

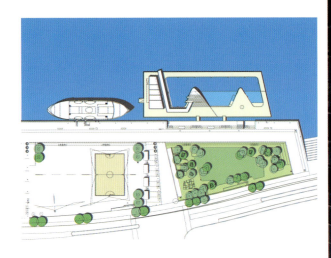

Architects: JWH arkitekter
Structural engineers: Ramboll
Location: Jomfru Ane Park 6, 9000 Aalborg, Denmark
Completion: 2012
Client: Municipality of Aalborg
Gross floor area: 2,000 sqm
Function: leisure
Built on water: floats on concrete pontoons

"Aqua sanat – water heals." – Sebastian Kneipp, German priest and founder of the naturopathic medicine movement

Architects: Schenk+Waiblinger Architekten
Location: Am Inselpark, Hamburg-Wilhelmsburg, Germany
Completion: 2013
Client: Hochtief Solutions AG formart, Hamburg
Gross floor area: 12,400 sqm
Function: living
Built on water: pile dwelling

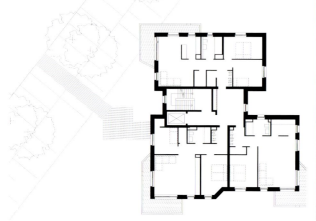

IBA WATERHOUSES
HAMBURG, GERMANY

This residential complex comprises five volumes: the nine-story WaterTower with 22 residential units and four TriplexHouses, each with three multi-story apartments with their own entrances. The entire complex is built on water and has a distinctive maritime flair. The Triplex apartments have direct access to the water. The Boathouse, a large communal space, is located at water level in the WaterTower and offers residents their own private access to the water. All buildings meet Passive House standards. Each resident can individually control their own energy use and gain. Thermal activation creates comfortable room conditions. The building received DGNB Gold certification from the German Sustainable Building Council.

"Water is a cultural medium. And it is our source: We ourselves consist of over 70 percent water." – Fabrizio Plessi, Italian artist

Architects: oth_architecten
Initiative and design: Trude Hooykaas
Structural engineers: Aronsohn raadgevende ingenieurs, Rotterdam
Location: Kraanspoor 12–58, IJ shore, North-Amsterdam, The Netherlands
Completion: 2007
Client: ING Real Estate Development
Gross floor area: 12, 500 sqm
Function: office
Built on water: rests on concrete supports

KRAANSPOOR
AMSTERDAM, THE NETHERLANDS

Kraanspoor is a lightweight, three-story office building, above a concrete crane-way of the former shipyard NDSM (Nederlandsche Dok en Scheepsbouw Maatschappij) in Amsterdam. The industrial heritage colossus dating back to 1952, has a length of 270 meters, a height of 13.5 meters and a width of 8.7 meters. Kraanspoor floats three meters above its concrete foundation, respecting the original architecture. The construction extends more on the waterside than on the shore side – due to the former cranes, which used to arch over this side of the waterfront. In the heart of the original concrete structure, immediately beneath the new office building, extensive archive space was created.

"The least movement is of importance to all nature. The entire ocean is affected by a pebble." — Blaise Pascal, French philosopher

This building system is designed so that the dyke houses, which are cantilevered on piles over the embankment, have similar structural detailing to the floating houses. The floating houses are supported on concrete tanks, which are submerged in water to a depth equal to half a story in height. The flotation tank doubles as a basement, and can be used for living space or for bedrooms. Each house is surrounded by a boardwalk which slopes down towards the water. With some 100 homes per hectare, the density of the neighborhood is comparable to that of the famous Jordaan district in central Amsterdam. The main urban design challenge was to give the water its due prominence as a distinctive feature of the neighborhood.

FLOATING HOUSES
AMSTERDAM, THE NETHERLANDS

Architects: Architectenbureau Marlies Rohmer
Structural engineers: Van der Vorm Engineering, ABC Arkenbouw, Kingma Bouw, Lelystad
Location: Waterbuurt-West, IJburg, Amsterdam, The Netherlands
Completion: 2011
Client: Ontwikkelingscombinatie Waterbuurt West & Woningstichting Eigen Haard
Gross floor area: 10,625 sqm
Function: living
Built on water: floats on a concrete tank

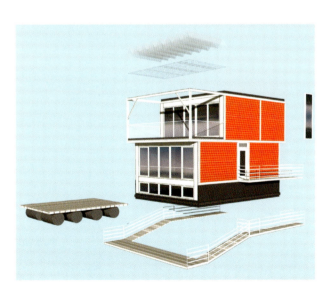

242 << 243

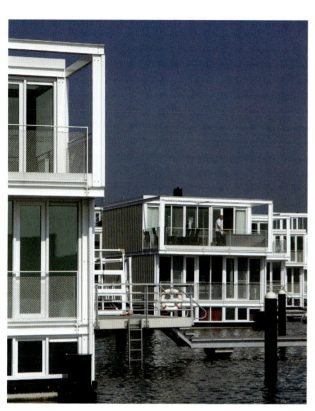

"Life originated in the sea, and about eighty percent of it is still there." – Isaac Aasimov, Russian-American bio-chemist and author

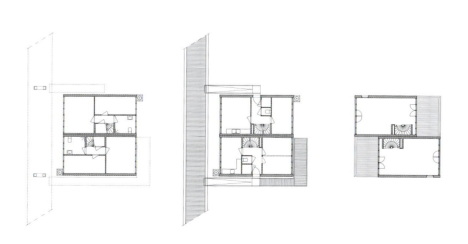

The Citadel is a floating apartment complex comprising 60 units. This is the first floating development with more than 30 units/4,047 square meters of water. The relatively high density leaves more open water surrounding the building. The floating complex is a composition of 180 modular elements arranged around a courtyard on top of a floating concrete caisson foundation. All the apartments have a view of the water and most have a berth for a small boat. The Citadel will be connected to the shore by means of a floating road. The building has a lightweight construction but special care has been given to sound insulation. The same standards have been met as for a normal land-based dwelling.

CITADEL
WESTLAND
THE NETHERLANDS

Architects: Waterstudio.NL
Location: Westland, The Netherlands
Completion: future
Client: ONW OPP/BNG
Gross floor area: 26,114 sqm
Function: living
Built on water: floats on hollow concrete foundation

The cohesive forces between liquid molecules are responsible for the phenomenon known as surface tension.

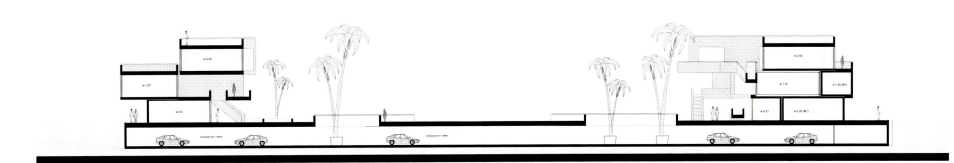

Water is one of the most vital elements for creating lively cities, but water is also one of our most precious and vulnerable elements. Like everywhere in the world, the quality and amount of water in Xinjin are under pressure. This master plan creates three new urban areas for Xinjin, that are attractive to live and work in, offer possibilities for commercial development and leisure activities and create space for water. The islands will be surrounded with a combination of dykes and in between levels. This soft boundary forms additional green areas in the dry season, and a close connection to the river when the water is higher. Each island can have its own unique identity, supporting the diverse public and commercial program, mixed with housing.

XINJIN WATER CITY
CHENGDU, CHINA

Achitects: MVRDV
Water management advisors: Arcadis, Rotterdam
Completion: ongoing
Client: Chengdu Life-City Investment
Gross floor area: 5,000,000 sqm
Function: living
Built on water: supported by platform resting on riverbed

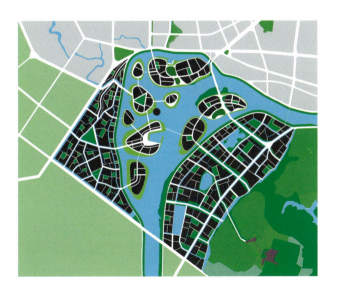

The Sun newspaper reported that the British Queen keeps a yellow plastic duck with an inflatable crown in her bath.

Architects: Büro Ole Scheeren
Location: Nai Pi Lae Lagoon, Kudu Island, Thailand
Completion: 2012
Client: community of Yao Noi
Gross floor area: 250 sqm
Function: leisure
Built on water: floats on wooden construction based on floating lobster farms

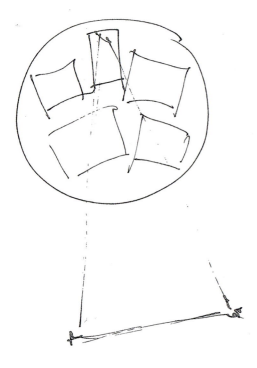

ARCHIPELAGO CINEMA
KUDU ISLAND, THAILAND

Designed by German-born and Beijing-based architect Ole Scheeren, Archipelago Cinema is an auditorium raft designed to float on the sea, premiered at the inaugural edition of the Film on the Rocks Yao Noi Festival, curated by Apichatpong Weerasethakul and Tilda Swinton. The Festival is set to become an annual meeting place for art and film. The raft is built out of recycled materials as a series of individual modules to allow for flexibility for its future use. Subsequent to a journey that will see the raft travel to further places as an auditorium for other film screenings on water, it will eventually return to the island and be donated to its actual builders, the community of Yao Noi, as its own playground and stage in the ocean.

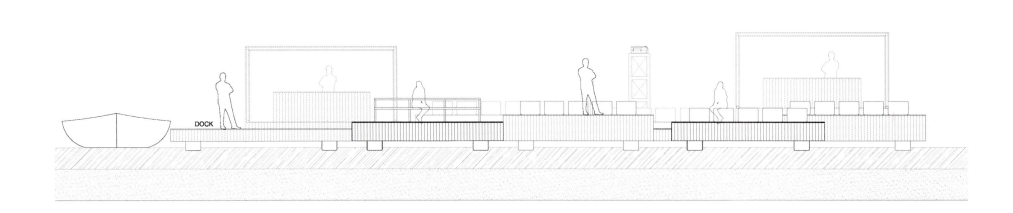

"We never know the worth of water till the well is dry." –
Thomas Fuller, English historian

Architects: Waterstudio.NL
Developers: Dutch Docklands Maldives
Location: North Malé Atoll, Maldives
Completion: future
Client: Dutch Docklands Maldives
Gross floor area: 55,170 sqm
Function: living
Built on water: floats on concrete foundation

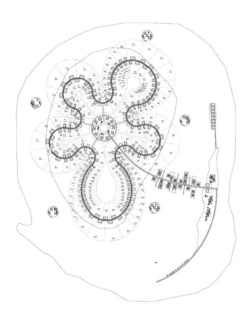

THE OCEAN FLOWER
MALÉ ATOLL, MALDIVES

The Ocean Flower comprises 185 luxury floating villas, each with direct access to a breathtaking turquoise lagoon with untouched corals and pristine beaches; all just 15 minutes away from the airport. These unique villas are being developed by Dutch Docklands Maldives, a joint venture company with the Maldives government and an exclusive affiliate of Christie's International Real Estate. The company offers clients the chance to own a villa but to rent it out as a five star establishment when it is not being used. The resort operator provides security, confidence and peace of mind and a lifestyle that has only been dreamed of until now.

"My soul is full of longing
for the secret of the sea,
and the heart of the great ocean
sends a thrilling pulse through me." –
Henry Wadsworth Longfellow, American writer

Architects: H&P Architects
Location: Cau Dien Town, Tu Liem District, Han Noi, Vietnam
Completion: 2013
Client: Hoang van Duy
Gross floor area: 44 sqm
Function: living, medical, office
Built on water: can float in flood water

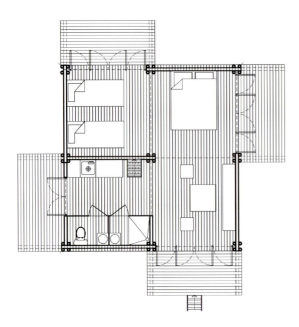

BLOOMING BAMBOO HOUSE
HA NOI, VIETNAM

The climate in Vietnam can be tough and a range of natural phenomena regularly causes catastrophes. The damage caused by drought, floods, landslides and storms can have a significant impact on communities and the need to find a quick and affordable housing solution is of paramount importance. The Blooming Bamboo House seeks to provide easy to assemble homes strong enough to withstand certain weather conditions, such as floods. When the area is flooded, the house can float on the water, withstanding floods of over one meter. The space is multi-functional and can be used as a home, classroom, medical center, or community center. It can be built by the users, without any need for expert assistance and can be constructed within a time frame of just 25 days.

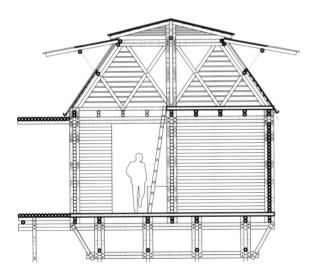

"The sea is everything. It covers seven tenths of the terrestrial globe. Its breath is pure and healthy. It is an immense desert, where man is never lonely, for he feels life stirring on all sides." – Jules Verne, French author

Architects: Jakob+MacFarlane
Location: Quai Rambaud, Lyon Confluence, Lyon, France
Completion: 2010
Client: Rhône Saône Développement / Cardinal
Gross floor area: 6,300 sqm
Function: culture
Built on water: supported by platform beneath

CUBE ORANGE
LYON, FRANCE

The ambition of the urban planning project for the old harbor zone, developed by VNF in partnership with Caisse des Dépôts and Sem Lyon Confluence, was to revamp the docks of Lyon on the riverside, bringing together architecture and a cultural and commercial program. The project is designed as a simple orthogonal cube into which a giant hole is carved, responding to necessities of light, air movement and views. This hole creates a void, piercing the building horizontally from the riverside inwards and upwards through the roof terrace. The cube's light façade, with seemingly random openings, is completed by another façade, pierced with pixilated patterns that accompany the movement of the river.

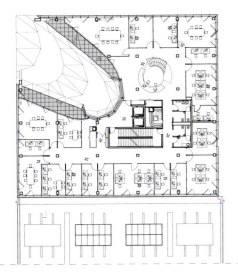

"A quotation in the right place at the right time is like water in a desert" – Vikrant Parsai, Indian author

INDEX

5468796 ARCHITECTURE www.5468796.ca — 116	**JAMES CARPENTER DESIGN ASSOCIATES** www.jcdainc.com — 144	**THOMAS FREIWALD** www.thomasfreiwald.com — 48
+31ARCHITECTS www.plus31architects.com — 66, 84	**JOANNA BOREK-CLEMENT** www.joannaborekclement.com — 152	**H ARCHITECTURE** www.h-architecture.com — 220
ERKKO AARTI, ARTO OLLILA & MIKKI RISTOLA (AOR) www.aor.fi — 124	**CONCRETE** www.concreteamsterdam.nl — 42	**HAEAHN ARCHITECTURE** www.haeahn.com — 220
ATTIKA ARCHITEKTEN www.attika.nl — 100, 138	**DFZ ARCHITEKTEN** www.dfz-architekten.de — 36	**H&P ARCHITECTS** www.hpa.vn — 262
BACA ARCHITECTS www.baca.uk.com — 70, 88, 112	**DREISSEN ARCHITECTS** www.dreissenarchitecten.nl — 106	**HWCD** www.h-w-c-d.com — 164
BAUBÜRO.EINS www.bbeins.com — 58	**DUGGAN MORRIS ARCHITECTS** www.dugganmorrisarchitects.com — 174	**JAGER JANSSEN ARCHITECTS** www.jagerjanssen.nl — 106
BAUMHAUER EICHLER ARCHITEKTEN www.baumhauer-architekten.de — 54, 134	**DUTCH DOCKLANDS INTERNATIONAL** www.dutchdocklands.com — 216	**JAKOB+MACFARLANE** www.jakobmacfarlane.com — 168, 266
MARIJN BEIJE www.marijnbeije.nl — 156	**DUTCH DOCKLANDS MALDIVES** www.dutchdocklands.com — 258	**JWH ARKITEKTER** www.jwh.dk — 228
BÜRO13 ARCHITEKTEN www.buero13.de — 62	**AFM ARCHITEKTEN MARTIN FÖRSTER** www.architekten-mf.de — 96	**MARTINOFF ARCHITEKTEN** www.martinoff-architekten.de — 192

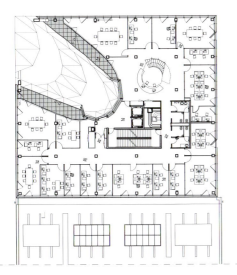

"A quotation in the right place at the right time is like water in a desert" – Vikrant Parsai, Indian author

INDEX

5468796 ARCHITECTURE www.5468796.ca — 116	JAMES CARPENTER DESIGN ASSOCIATES www.jcdainc.com — 144	THOMAS FREIWALD www.thomasfreiwald.com — 48
+31ARCHITECTS www.plus31architects.com — 66, 84	JOANNA BOREK-CLEMENT www.joannaborekclement.com — 152	H ARCHITECTURE www.h-architecture.com — 220
ERKKO AARTI, ARTO OLLILA & MIKKI RISTOLA (AOR) www.aor.fi — 124	CONCRETE www.concreteamsterdam.nl — 42	HAEAHN ARCHITECTURE www.haeahn.com — 220
ATTIKA ARCHITEKTEN www.attika.nl — 100, 138	DFZ ARCHITEKTEN www.dfz-architekten.de — 36	H&P ARCHITECTS www.hpa.vn — 262
BACA ARCHITECTS www.baca.uk.com — 70, 88, 112	DREISSEN ARCHITECTS www.dreissenarchitecten.nl — 106	HWCD www.h-w-c-d.com — 164
BAUBÜRO.EINS www.bbeins.com — 58	DUGGAN MORRIS ARCHITECTS www.dugganmorrisarchitects.com — 174	JAGER JANSSEN ARCHITECTS www.jagerjanssen.nl — 106
BAUMHAUER EICHLER ARCHITEKTEN www.baumhauer-architekten.de — 54, 134	DUTCH DOCKLANDS INTERNATIONAL www.dutchdocklands.com — 216	JAKOB+MACFARLANE www.jakobmacfarlane.com — 168, 266
MARIJN BEIJE www.marijnbeije.nl — 156	DUTCH DOCKLANDS MALDIVES www.dutchdocklands.com — 258	JWH ARKITEKTER www.jwh.dk — 228
BÜRO13 ARCHITEKTEN www.buero13.de — 62	AFM ARCHITEKTEN MARTIN FÖRSTER www.architekten-mf.de — 96	MARTINOFF ARCHITEKTEN www.martinoff-architekten.de — 192

C.F. MØLLER ARCHITECTS www.cfmoller.com — 74	DENIS OUDENDIJK www.refunc.nl — 148	SLO ARCHITECTURE www.sloarchitecture.com — 212
MONO ARCHITEKTEN www.monoarchitekten.de — 58	PAD STUDIO www.PADstudio.co.uk — 8	SPRENGER VON DER LIPPE ARCHITEKTEN www.sprengervonderlippe.de — 130
MVRDV www.mvrdv.nl — 160, 250	REMISTUDIO www.remistudio.ru — 26	TUN-ARCHITEKTUR www.tun-architektur.de — 14
ROBERT NEBOLON ARCHITECTS www.RNarchitect.com — 182, 224	MAREK ŘÍDKÝ — 18	WATERSTUDIO.NL www.rosalespartners.com — 22, 112, 216, 246, 258
NIO ARCHITECTEN www.nio.nl — 188, 196	ARCHITECTENBUREAU MARLIES ROHMER www.rohmer.nl — 240	WHIM ARCHITECTURE www.whim.nl — 208
ONL [OOSTERHUIS_LÉNÁRD] www.oosterhuis.nl — 178	ROST.NIDEREHE ARCHITEKTEN I INGENIEURE www.rost-niderehe.de — 204	MONIKA WIERZBA www.mwierzba.carbonmade.com — 120
OPEN DEVELOPMENT www.opendevelopment.nl — 92	BÜRO OLE SCHEEREN www.buro-os.com — 254	WILK-SALINAS ARCHITEKTEN www.wilk-salinas.com — 48
ROBERT HARVEY OSHATZ, ARCHITECT www.oshatz.com — 30	SCHENK+WAIBLINGER ARCHITEKTEN www.schenk-waiblinger.de — 232	WXCA www.wxca.pl — 200
OTH_ARCHITECTEN www.oth.nl — 92, 236	HAN SLAWIK ARCHITEKT / ARCHITECH www.architech.pro — 78	

PICTURE CREDITS

ABC Arkenbouw 105 a. · Roos Aldershoff 240 · Archimage Architectural Photography/Meike Hansen 14–16, 17 a. r. · Archimages 130–133 · Arena Berlin 48, 51 · Iwan Baan 66–69 · Martine Berendsen 139, 142 · Nicolas Borel 168–173, 266–269 · Doug Bruce 256 a. l. · Christiaan de Bruijne 236–239 · Marcel van der Burg 244 b. · Max Creaasy 124, 125, 128 · N.Dudda / T. Müller 17 a. l · Concrete/Photographer: Jim Ellam 46, 47 b. · Courtesy of Film on the Rocks, Yao Noi Foundation 256 a. r., 257 · Doan Thanh Ha 262–265 · Piyatat Hemmatat, 254, 255 · Jack Hobhouse, London 174–177 · Bart von Hoek 100–104, 105 b., 140, 141 r., 143 a. · Concrete/Photographer: Ewout Huibers 42 r., 43–45, 47 a. · idem 106, 107, 109–111 · Stanisław Kempa, Piotr Skrzycki, Monika Wierzba 120–123 · Sebastian Kochel 200, 201 · Christian van der Kooy 149, 150 · Luuk Kramer 241, 242, 243 l. 244 a. 245 · Jens Kroell, Hamburg 204, 205 b. l. 206, 207 · Martin Kunze, Hamburg 192–195 · Yong Kwan Kim, Seoul, KR 220–223 · David Duncan Livingston 226, 227 · Rop van Loenhout 141 l., 143 b. · Floris Lok 243 r. · John Lewis Marshall, Amsterdam 108 · Martrade Immobilien Dusseldorf 134, 135, 137 · Matthew Millman 182–187 · Max Missal, Hamburg · Kleusberg GmbH & Co.KG/ Photographer: Rüdiger Mosler 78–83 · Anke Müllerklein 232–235 · Cameron Neilson 30–35 · Rickerd van der Plas 138 · Jock Pottle 147 b. · Marek Řídký 18–21 · Nigel Rigden 8–13 · Amelie Rost, Hamburg 205 a., b. r. · Thorsten Seidel 49, 50, 52, 53 · Arek Seredyn, NIO architecten 188–191 · Hagen Stier, Hamburg 37–41 · Andreas Symietz, New York 212, 213, 214 a., 215 · T+E Shanghai 164–167 · Charles Tang, New York 214 b. · Wieland&Gouwens, Rotterdam 160–163 · Xocolatl / Wikimedia Commons

Cover front: Iwan Baan
Cover back (from left to right, from above to below):
Concrete/Ewout Huibers; Remistudio; Hagen Stier, Hamburg; Bart von Hoek; Nigel Rigden

IMPRINT

The Deutsche Nationalbibliothek lists this publication in the Deutsche Nationalbibliografie; detailed bibliographic data are available in the Internet at http://dnb.dnb.de

ISBN 978-3-03768-178-7
© 2015 by Braun Publishing AG
www.braun-publishing.ch

The work is copyright protected. Any use outside of the close boundaries of the copyright law, which has not been granted permission by the publisher, is unauthorized and liable for prosecution. This especially applies to duplications, translations, microfilming, and any saving or processing in electronic systems.

1st edition 2015

Selection of projects: Editorial office van Uffelen
Editorial staff and layout: Lisa Rogers, Chris van Uffelen
Graphic concept: Michaela Prinz, Berlin
Reproduction: Bild1Druck GmbH, Berlin

All of the information in this volume has been compiled to the best of the editor's knowledge. It is based on the information provided to the publisher by the architects' and designers' offices and excludes any liability. The publisher assumes no responsibility for its accuracy or completeness as well as copyright discrepancies and refers to the specified sources (architects' and designers' offices). All rights to the photographs are property of the photographer (please refer to the picture credits).